Principles & Practices of Quality Assurance

SAGE was founded in 1965 by Sara Miller McCune to support the dissemination of usable knowledge by publishing innovative and high-quality research and teaching content. Today, we publish over 900 journals, including those of more than 400 learned societies, more than 800 new books per year, and a growing range of library products including archives, data, case studies, reports, and video. SAGE remains majority-owned by our founder, and after Sara's lifetime will become owned by a charitable trust that secures our continued independence.

Los Angeles | London | New Delhi | Singapore | Washington DC | Melbourne

Principles & Practices of Quality Assurance

A guide for internal and external quality assurers
in the FE and skills sector

Ann Gravells

Learning Matters
An imprint of SAGE Publications Ltd
1 Oliver's Yard
55 City Road
London EC1Y 1SP

SAGE Publications Inc.
2455 Teller Road
Thousand Oaks, California 91320

SAGE Publications India Pvt Ltd
B 1/I 1 Mohan Cooperative Industrial Area
Mathura Road
New Delhi 110 044

SAGE Publications Asia-Pacific Pte Ltd
3 Church Street
#10-04 Samsung Hub
Singapore 049483

Editor: Amy Thornton
Production Controller: Chris Marke
Project Management: Deer Park Productions,
Tavistock, Devon
Marketing Manager: Lorna Patkai
Cover Design: Wendy Scott
Typeset by: C&M Digitals (P) Ltd, Chennai, India
Printed and bound by CPI Group (UK) Ltd,
Croydon, CR0 4YY

Library of Congress Control Number: 2016940062

British Library Cataloguing in Publication Data

A catalogue record for this book is available from the
British Library.

ISBN: 978-1-4739-7342-8
ISBN: 978-1-4739-7341-1 (hbk)

At SAGE we take sustainability seriously. Most of our products are printed in the UK using FSC papers and boards.
When we print overseas we ensure sustainable papers are used as measured by the PREPS grading system.
We undertake an annual audit to monitor our sustainability.

CONTENTS

ACKNOWLEDGEMENTS

I would like to give a special thanks to the following people who have helped me with the production of this book. They have freely given their time, knowledge and advice, which has resulted in some excellent contributions to the content.

Alison Quilliam – Group Quality Manager at The LTE Group

Andrew Walker – Managing Director at Training Qualifications UK (TQUK)

Angela O'Leary – Lead TAQA Trainer, Assessor and IQA at Leeds College of Building

Bronia Davis – Co-ordinator of Internal Quality Assurance, Training & Development at East Riding College

Joey Greenwood – Training Director at Smart Training Solutions (UK) Limited

Roisin Kelly – Skills Development Coordinator at NICVA

Sharron Carlill, Assistant Principal (Quality and Compliance) at The White Rose Beauty Colleges

Sharon Haggerty – Freelance Training Practitioner

Sue Lloyd – Quality Improvement Co-ordinator at North Lindsey College

I would like to thank my Senior Commissioning Editor (Education), Amy Thornton, and my Development Editor, Jennifer Clark, for their continued support and guidance.

Every effort has been made to trace the copyright holders and to obtain their permission for the use of copyright material. The publisher and author will gladly receive any information enabling them to rectify any error or omission in subsequent editions.

Ann Gravells
www.anngravells.com

Ann has been teaching and assessing in the further education and skills sector since 1983. She is a director of her own company Ann Gravells Ltd, an educational consultancy based in East Yorkshire. She specialises in teaching, training, assessment and quality assurance for the further education and skills sector.

Ann holds a Masters in Educational Management, a PGCE, a Degree in Education and a City & Guilds Medal of Excellence for teaching. She is a Fellow of the Society for Education and Training, and holds QTLS status.

Ann creates resources for teachers and learners such as PowerPoints and handouts for the assessment, quality assurance and teacher training qualifications. These are available via her website: www.anngravells.com.

Ann has worked for several awarding organisations producing qualification guidance, policies and procedures, and carrying out external quality assurance of teaching, assessment and quality assurance qualifications.

She is currently a consultant to the University of Cambridge's Institute of Continuing Education and to Training Qualifications UK (TQUK), the UK's fastest growing awarding organisation which specialises in Education and Training, Assessment and Quality Assurance qualifications.

The author welcomes any comments from readers; please contact her via her website: www.anngravells.com

Ann Gravells is the author of the following books, many of which are in new editions (listed here in alphabetical order). Details can be found on her website: www.anngravells.com

Achieving your Assessment and Quality Assurance Units (TAQA)

Delivering Employability Skills in the Lifelong Learning Sector

Passing Assessments for the Award in Education and Training

Passing PTLLS Assessments

Preparing to Teach in the Lifelong Learning Sector

Principles and Practices of Assessment

Principles and Practices of Quality Assurance

The Award in Education and Training

What is Teaching in the Lifelong Learning Sector?

She is co-author of:

Equality and Diversity in the Lifelong Learning Sector

Passing CTLLS Assessments

Planning and Enabling Learning in the Lifelong Learning Sector

The Best Vocational Trainer's Guide

The Certificate in Education and Training

Passing Assessments for the Certificate in Education and Training

She has edited:

Study Skills for PTLLS

INTRODUCTION

Congratulations on purchasing this book. Whether you are a new or an experienced internal or external quality assurer, this book will guide you through the terminology, principles and practices to enable you to become a quality assurer, improve your role and/or work towards a relevant qualification if necessary. The book is also applicable to anyone taking quality assurance units which form part of the teacher training or learning and development qualifications.

You might benefit from reading the book *Principles and Practices of Assessment* (2016) by the same author, to refresh your knowledge and practice of assessment.

Your role as a quality assurer might differ depending upon the type of employment contract you have, for example, if you are full time, part time, freelance or self-employed. As you work through the book, some aspects might therefore not apply to you.

Some aspects of this book are also in the book *Achieving your Assessment and Quality Assurance Units (TAQA)* (2014) by the same author.

Due to the terminology used throughout the further education and skills sector, you will find lots of abbreviations and acronyms within the book. A list of the most commonly used ones can be found in the Appendix.

The term *learner* is used throughout the book to denote anyone taking a qualification or programme of learning and who might not necessarily call themselves a learner, for example, an employee or apprentice in the workplace. The term *assessor* is used for anyone who assesses; however, the assessor might also be a teacher, trainer, coach or mentor.

There are activities and examples within each chapter which will assist your understanding of quality assurance principles and practices. At the end of each section within the chapter is an extension activity to stretch and challenge your learning further. Completing the activities will help you put theory into practice and contribute towards your continuing professional development.

Throughout the chapters there are examples of completed templates that could be used or adapted for quality assurance purposes. However, do check with the organisation you are working for, in case they have particular documents they require you to use. For the purpose of future proofing the book, a year has not been added to any dates within the tables and checklists in the chapters. You should add the year as well as the day and month to create a full audit trail.

At the end of each chapter is a list of relevant textbooks and websites enabling you to research topics further. As these are often revised and updated, some might become unavailable over time.

The index at the back of the book will help you quickly locate useful topics within the book.

1 PRINCIPLES OF INTERNAL QUALITY ASSURANCE

Introduction

Quality assurance is the general term used for making sure things are the best they can be. It should be a continual process with the aim of maintaining and improving the products and services offered in an organisation. Internal quality assurance relates to the monitoring of all the teaching, learning and assessment activities which learners or employees will undertake, and forms part of an organisation's quality cycle. If internal quality assurance does not take place, there are risks to the accuracy, consistency and fairness of teaching, learning and assessment practice.

This chapter will explore the role of an internal quality assurer, along with the concepts and principles which underpin it.

This chapter will cover the following topics:

- Quality assurance
- Roles and responsibilities of an internal quality assurer
- Concepts and principles of internal quality assurance
- Risk management
- Maintaining and improving the quality of assessment
- The role of technology in internal quality assurance

Quality assurance

Quality *assurance* can be defined as a system which guarantees the quality of the products or services offered. Within education and training contexts, the product is what a learner is hoping to achieve (i.e. a qualification or to maintain their standards of working). The service is the support the learner or employee receives from their supervisor, trainer, assessor and others. The quality assurance process should seek to avoid problems, stabilise, and improve the products and services offered to learners. It can be thought of as a *proactive* system to resolve any issues as they occur. This is in contrast to quality *control* which seeks to find problems, and is usually *reactive* after the event. A good quality system will have structures in place to enable situations to be dealt with as they happen, rather than afterwards when it might be too late. Quality *improvement* will be achieved by monitoring and evaluating the quality system. *Internal* quality assurance (IQA) will form part of the overall quality assurance system in an organisation.

An education or training organisation might follow a *quality cycle* to ensure all aspects of their provision are planned, delivered, assessed and evaluated. Following the aspects in the quality cycle should ensure the best provision of all the products and services offered. An organisation's quality cycle could involve the aspects as in Figure 1.1 on page 4 which would be overseen by a quality manager. However, not all of these aspects might occur in the organisation you are working in, but it would be worth finding out which do. If carried out correctly, the quality process should safeguard the credibility of the qualification, programmes or standards offered. It's much more difficult to audit poor quality out of a system than to build good quality in.

Quality assurance is not just a means to an end or something that has to be done because someone says so. There should be a purpose for each aspect of the process which will ensure everything runs smoothly. Aspects should be evaluated to enable changes and/or improvements to take place as necessary.

A quality cycle will often cover activities which are in addition to, or which complement, those which might occur as a part of the internal quality assurance process. The IQA process is explained after Figure 1.1 and covers the activities you are more likely to carry out as part of your role. However, all aspects of the quality cycle are relevant to the way an organisation which offers education and training should operate. For example, if a budget was not agreed prior to a programme commencing, money could run out part way through and the training would be postponed or cancelled.

Activity

Look at the quality cycle in Figure 1.1 on page 4. Which aspects are carried out in your organisation? What other aspects might take place and why? Will you be required to be involved with any, and if so, how will this impact upon your IQA role?

If you have access to the internet, carry out a search to find out the differences between quality assurance and internal quality assurance.

Internal quality assurance

Internal quality assurance (IQA) relates to the monitoring of all the teaching, learning and assessment activities which learners or employees will undertake. Monitoring should form part of an organisation's quality cycle, as in Figure 1.1. If the organisation offers several programmes, there might be quite a few staff carrying out the IQA role for the different subjects offered, for example, hairdressing, retail and customer service. If so, they will need to liaise with each other to standardise their practice and ensure a consistent service is given to assessors and learners.

IQA monitoring activities will help you find out if your trainers and assessors are performing their roles as they should, and to give them advice and support as necessary. Your role might be to monitor the whole process from when a learner commences to when they

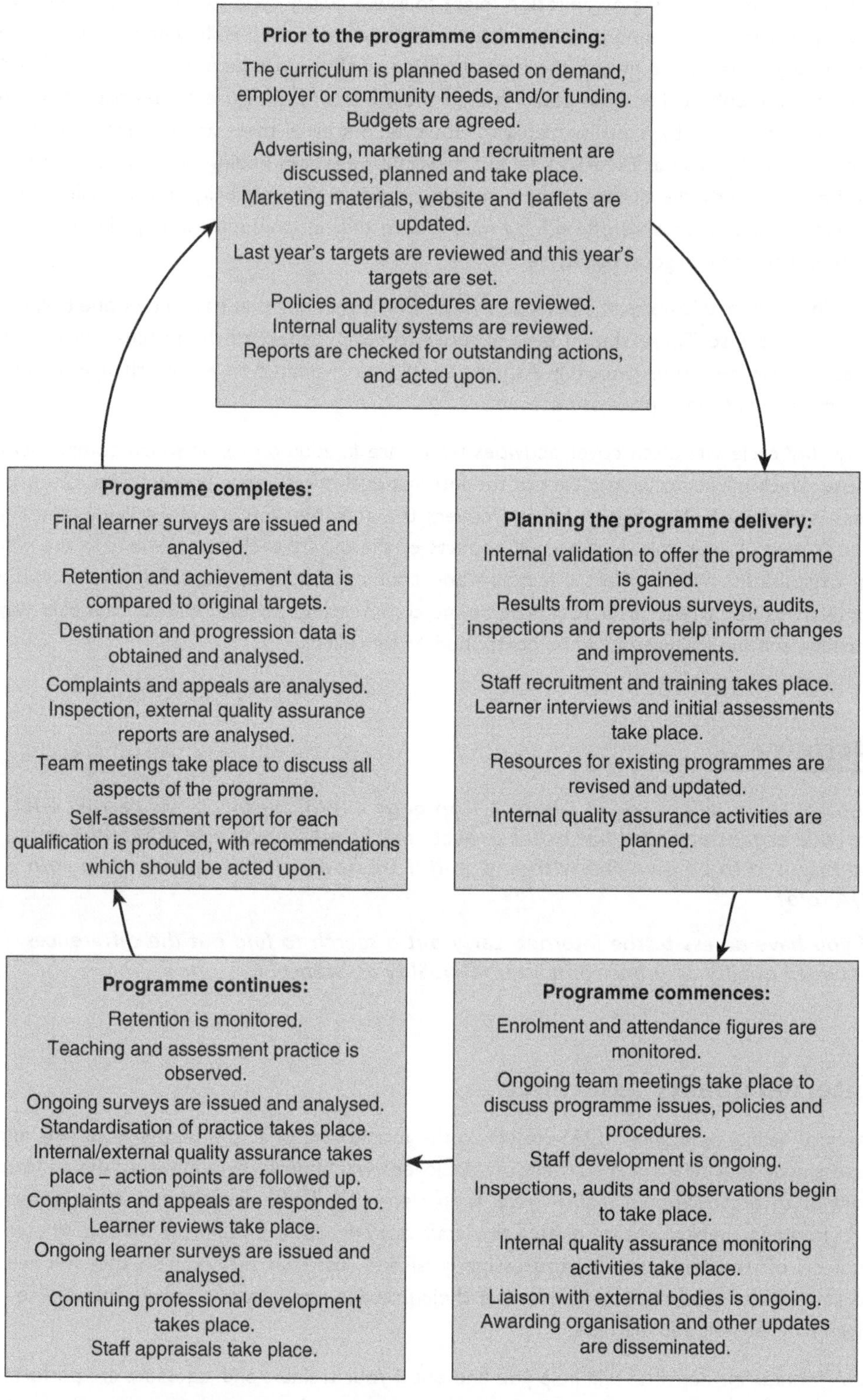

Prior to the programme commencing:

The curriculum is planned based on demand, employer or community needs, and/or funding. Budgets are agreed.

Advertising, marketing and recruitment are discussed, planned and take place.
Marketing materials, website and leaflets are updated.

Last year's targets are reviewed and this year's targets are set.
Policies and procedures are reviewed.
Internal quality systems are reviewed.
Reports are checked for outstanding actions, and acted upon.

Programme completes:

Final learner surveys are issued and analysed.

Retention and achievement data is compared to original targets.

Destination and progression data is obtained and analysed.

Complaints and appeals are analysed.
Inspection, external quality assurance reports are analysed.

Team meetings take place to discuss all aspects of the programme.

Self-assessment report for each qualification is produced, with recommendations which should be acted upon.

Planning the programme delivery:

Internal validation to offer the programme is gained.
Results from previous surveys, audits, inspections and reports help inform changes and improvements.

Staff recruitment and training takes place.
Learner interviews and initial assessments take place.

Resources for existing programmes are revised and updated.

Internal quality assurance activities are planned.

Programme continues:

Retention is monitored.

Teaching and assessment practice is observed.

Ongoing surveys are issued and analysed.
Standardisation of practice takes place.
Internal/external quality assurance takes place – action points are followed up.
Complaints and appeals are responded to.
Learner reviews take place.
Ongoing learner surveys are issued and analysed.
Continuing professional development takes place.
Staff appraisals take place.

Programme commences:

Enrolment and attendance figures are monitored.
Ongoing team meetings take place to discuss programme issues, policies and procedures.
Staff development is ongoing.
Inspections, audits and observations begin to take place.
Internal quality assurance monitoring activities take place.
Liaison with external bodies is ongoing.
Awarding organisation and other updates are disseminated.

Figure 1.1 An example of an organisation's quality cycle

finish, i.e. the full learner journey. IQA activities can also take place prior to the learner commencing, i.e. by monitoring the application and interview process, to after they have left, i.e. following up on progression opportunities. Internal quality assurance should not be in isolation from other aspects of teaching, learning and assessment. It's not something that is added on to the end of a qualification or a programme of learning. It should be carried out on an ongoing basis, with a view to making improvements, or keeping the status quo if everything is satisfactory. Records must be maintained of all monitoring activities to prove they actually took place and to facilitate improvements as necessary.

Example

A training organisation monitors the practice of their trainers and assessors by observing them with their learners. As the observations are ongoing, any issues can be identified and resolved straight away. If the observations did not take place, problems might not be identified, learners would be dissatisfied and the organisation's high standards would not be maintained.

There are other terms used for internal quality assurance which you might come across, for example, internal *verification* and internal *moderation*. Internal *verification* can be considered a lesser version of internal quality assurance and is often for vocational subjects. The process will seek to sample the work of assessors over time, but not sample all aspects associated with the full learner journey. If there is a problem with the sample, the assessor will need to revisit it with their learners.

Internal *moderation* seeks to sample a proportion of assessed work for a particular aspect, and is often for non-vocational subjects. If the sample finds problems, then the assessor will need to revisit it with every learner, including those outside of the original sample. Internal moderation is sometimes referred to as double marking or re-assessment as it confirms the original assessment decisions were accurate (or not).

Verification and moderation tend to be *reactive* after an event, whereas internal quality assurance aims to be *proactive* throughout the learner journey to prevent problems occurring. You will need to find out if any aspect of your role will involve verification and/or moderation, or if it is just based on quality assurance monitoring.

Figure 1.2 on page 6 shows the different internal quality assurance models, along with some of the activities which might be performed. Your role might include some or all of them depending upon the products or services offered. The activities in column three will be explained in detail in Chapter 2.

Often, internal quality assurers are also experienced trainers and assessors in the subject area they are quality assuring. For example, if the subject area is motor vehicle maintenance, they should not be internally quality assuring other subjects they are not experienced in, for example, horticulture. The IQA process might be the same for each subject, but the internal quality assurer must be fully familiar with what is being assessed to

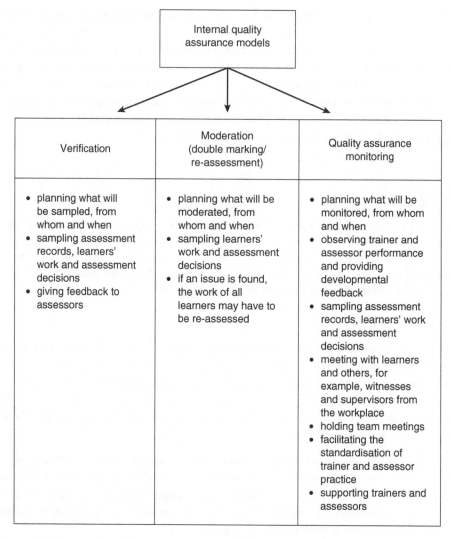

Figure 1.2 Internal quality assurance models

make a valid and reliable decision. Valid means you are doing what you should, and reliable means you would get similar results each time you did it.

You might be quality assuring an *accredited* qualification or an *endorsed* programme of learning from an awarding organisation (AO). An AO is regulated in the United Kingdom by Ofqual in England, DCELLS in Wales, CCEA in Northern Ireland and SQA in Scotland (links to these are at the end of the chapter in the website list). A training organisation offering accredited qualifications or endorsed programmes of learning is often referred to as a 'centre'.

An accredited qualification can be offered by several awarding organisations and is recognised nationally. Endorsed qualifications have been specifically written by a centre, in conjunction with one awarding organisation, to meet the needs of particular employers or learners. However, it could be that you are quality assuring a *non-accredited* programme of learning. This means a qualification will not be issued via an awarding organisation, for example, if a learner is demonstrating their competence towards their role specification, a set of workplace standards or their job description. The learner will be proving their skills and knowledge rather than working towards the achievement of a qualification. However, some learners might be doing both, for example, if they are taking an apprenticeship programme.

You will need to read the *assessment strategy* which will be part of the qualification or pro-gramme specification. This will state whether you must also be a qualified trainer and/or assessor in the same subject that you will internally quality assure. It will also state if asses-sors and IQAs need relevant experience and/or qualifications in the particular subject learners are working towards. You should also be fully familiar with the content in the qualifi-cation specification, i.e. what is to be delivered and assessed, as well as relevant requirements, regulations and standards.

The internal quality assurance cycle

Depending upon the subject to be quality assured, an IQA cycle will usually be followed as in Figure 1.3. This example relates to the monitoring of the teaching, learning and assess-ment process. The results and findings from all monitoring activities should be discussed with relevant staff, and improvements planned and carried out as necessary.

Figure 1.3 Internal quality assurance (IQA) cycle

The IQA cycle can involve the following aspects:

- **Identify the product or service** – ascertain what is to be assessed and internally quality assured and why. For example, are learners working towards a qualification or a programme of learning, or are staff being observed performing their job roles? The criteria will need to be clear, i.e. units from a qualification (product) or the support the learner receives (service). Learners should be allocated to assessors in a fair way, for example, according to their location or workload.

- **Planning** – devise a sample plan to arrange what will be monitored, from whom and when. Plan the dates to observe trainer and assessor performance, hold team meetings and standardisation activities. Information will need to be obtained from assessors to assist the planning process, and risks taken into account such as assessor knowledge, qualifications and experience.

- **Activity** – carry out the IQA activities such as sampling learners' work, talking to learners, staff, supervisors and witnesses, observing trainer and assessor performance, sampling assessment records and decisions, and preparing for external quality assurance (EQA) visits. Activities also include holding meetings and standardisation activities, supporting and training relevant staff, and communicating with others involved in the assessment and IQA process. These will be explained in Chapter 2.

- **Decision and feedback** – make a judgement as to whether the trainer or assessor has performed satisfactorily and made valid and reliable decisions. Provide developmental feedback as to what was good or what could be improved. Agree action points if necessary and follow them up.

- **Evaluation** – carry out a review of the assessment and IQA process to determine what could be improved or done differently. Agree action plans if necessary; implement and follow up. Follow any action plans from external quality assurers or inspectors. Write self-assessment reports as necessary, to link in with the quality cycle.

Throughout the cycle, standardisation of practice between assessors and internal quality assurers should take place; this will help ensure the consistency and fairness of all decisions. Feedback should also be obtained from learners and others involved in the assessment process. Records must be maintained of all activities for audit requirements. If the qualification is accredited or endorsed by an awarding organisation, EQA will also take place. Please see Chapter 2 regarding preparing for an EQA visit, and Chapters 3 and 4 for information regarding the EQA process.

As an internal quality assurer only *samples* various activities, there is the possibility some aspects might be missed. Imagine this taking place in a bakery: the quality assurer would not sample every item by tasting each one made. They would only taste a sample from each baker, otherwise there would be nothing left to sell. There are often risks involved with sampling and these should be considered when planning what will be internally quality assured, when and how. For example, if an assessor is new to their role their work should be sampled more.

Extension activity

Look at the above bulleted list following Figure 1.3. Describe how each aspect of the cycle will impact upon your role as an IQA. Is there anything you are unsure of or missing from the cycle? As you progress through Chapters 1 and 2, it should all become clearer.

Roles and responsibilities of an internal quality assurer

Your main role will be to carry out the IQA process according to what is being assessed, for example, a qualification, programme of learning, work tasks, a job description or a role specification. You also need to follow all relevant regulations and guidelines. Your own roles and responsibilities may or may not include those in the IQA cycle in Figure 1.3. There might be a quality assurance manager in your organisation who will be responsible for some of the aspects. You therefore need to know what you are going to internally quality assure, when and where the monitoring activities will take place, and how you will go about them. If you are currently quality assuring, you will probably know this already; if you are new, you will need to find this out.

The type of employment contract you have and where you are based will have an influence on how often you see your assessors and the types of monitoring activities you will perform. This will differ if you are full time, part time, freelance or self-employed, and if you are based in the same location as your assessors or not. Your staff might have different types of contracts too and perhaps be working for more than one organisation. It is therefore important to find this out and ensure they are following the requirements of your organisation. Sometimes, if staff work for more than one organisation, the procedures and documents used will differ and the wrong ones might be used by accident. It's also possible that you might have to perform some tasks, for example, administration, in your own time.

Your roles and responsibilities might include (in alphabetical order):

- advising, supporting and providing developmental feedback to assessors

- documenting the internal quality assurance rationale and strategy, process and decisions

- ensuring trainers and assessors interpret, understand and consistently apply the relevant requirements, regulations and standards

- identifying issues and trends, for example, if several learners misinterpret the same topic

- interviewing learners, assessors and other relevant staff

- keeping up to date with changes and developments regarding relevant standards and qualifications

- leading standardisation activities to ensure the accuracy and consistency of assessment decisions between assessors

- monitoring and observing the full learner journey from commencement to completion

- planning and carrying out the sampling of assessed work

- reflecting on your role

- taking part in continuing professional development (CPD)

- working towards relevant IQA qualifications.

Internal quality assurers are often supervisors or managers and are naturally responsible for staff, systems and procedures. However, some are still working as assessors and performing both roles. That's absolutely fine as long as they don't IQA their own assessment decisions, as that would be a conflict of interest. Some smaller organisations might only have one assessor and one internal quality assurer, which again is fine providing they remain fully objective when carrying out their role. Some small teams, i.e. one assessor and one internal quality assurer, can swap roles and IQA each other's assessment decisions. Again, it's not a problem unless it's towards an accredited qualification and the awarding organisation deems it is. It could be considered a good way of standardising practice, as they will be monitoring each other regularly to ensure consistency.

Example

Angela is the IQA as well as the assessor for a qualification in fabrication and welding. There are 40 learners and one other assessor/IQA: Aleksy. Angela and Aleksy each have a case load of 20 learners and IQA the work of each other. They are therefore not quality assuring their own assessment decisions. This has been agreed with the awarding organisation and also helps to standardise their practice as they each see what the other is doing.

If there is a team of IQAs in your organisation and you are responsible for them, you will find Chapter 5 helpful.

Internal quality assurers might also be required to carry out other roles, some of which form part of the quality cycle in Figure 1.1 These might include (in alphabetical order):

- analyse enrolment, retention, achievement and progression data

- carry out a training needs' analysis with trainers and assessors

- compile self-assessment reports

- countersign other internal quality assurers' judgements

- deal with appeals and complaints

- design advertising and marketing materials

- design, issue and analyse questionnaires and surveys, and set action plans based on the findings

- ensure qualifications are fit for purpose and validated by the organisation

- ensure IQA strategies, policies and procedures are regularly reviewed

- facilitate appropriate staff development, training and CPD

- implement any action points from reports

- induct and mentor new staff, support existing staff and carry out staff appraisals

- interview new staff

- liaise with others involved in the IQA process, e.g. trainers, witnesses from the workplace, external quality assurers, administrative staff

- prepare agendas and chair meetings

- prepare for external inspections and visits from awarding organisation personnel

- provide statistics and reports to line managers

- register and certificate learners with an awarding organisation

- set targets and/or performance indicators.

Activity

If you are currently carrying out the IQA role, complete numbers 1 and 2 below. If not, just complete number 1.

1   *Obtain a copy of what is to be assessed and quality assured. For example, the qualification specification, programme, work tasks, role specification or job description. Have a look at the requirements for trainers, assessors and IQAs regarding qualifications and experience. Read through the content to ensure you are fully familiar with it.*

2   *Find out who the assessors are that you will be responsible for and obtain their contact details. If you are quality assuring an accredited or endorsed qualification, you will need to make sure the awarding organisation is kept up to date with changes to staffing details.*

Legislation

Legislation will differ depending upon the context and environment within which you will internally quality assure. You also might need to be aware of the requirements of external bodies and regulators such as Ofsted (in England) who inspect funded provision and Ofqual (in England) who regulate awarding organisations.

The following information was current at the time of writing; however, you are advised to check for any changes or updates, and whether they are applicable outside England.

Children Act 2004 will be applicable to you if you work with 14–19-year-olds, vulnerable adults or learners with special needs.

Copyright Designs and Patents Act 1988 relates to the copying, adapting and distributing of materials, which includes computer programs and materials found via the internet. Organisations can have a licence to enable the photocopying of small amounts from books or journals. All copies should have the source acknowledged.

Counter-Terrorism and Security Act 2015 will apply if you work with learners who are at risk of becoming radicalised. The Prevent Duty is part of this Act.

Data Protection Act 1998 made provision for the regulation of the processing of information relating to individuals, including the obtaining, holding, use or disclosure of such information. It was amended in 2003 to include electronic data.

The Equality Act 2010 replaced all previous anti-discrimination legislation and consolidated it into one Act (for England, Scotland and Wales). It provides rights for people not to be directly discriminated against or harassed because they have an association with a disabled person or because they are wrongly perceived as disabled.

Freedom of Information Act 2000 gives learners the opportunity to request to see the information public bodies hold about them.

Health and Safety at Work etc. Act 1974 imposes obligations on all staff within an organisation commensurate with their role and responsibility. Risk assessments should be carried out where necessary. In the event of an accident, particularly one resulting in death or serious injury, an investigation by the Health and Safety Executive may result in the prosecution of individuals found to be negligent as well as the organisation.

Rehabilitation of Offenders Act 1974 will be applicable if you work with ex-offenders.

Safeguarding Vulnerable Groups Act 2006 introduced a vetting and barring scheme to make decisions about who should be barred from working with children and vulnerable adults. Assessors may need to apply to the Disclosure and Barring Service (DBS) to have a criminal records' check. The purpose of the DBS is to help employers to prevent unsuitable people from working with children and vulnerable adults.

Welsh Language Act 1993 put the Welsh language on an equal footing with the English language in Wales, with regard to the public sector.

Regulatory requirements

Regulations are often called *rules* and they specify mandatory requirements that must be met. Public bodies, corporations, agencies and organisations create regulatory requirements which must be followed if they are applicable to the job role. For example, in education, one of the regulators is Ofqual who regulate qualifications, examinations and assessments in England. Ofqual gives formal recognition to awarding organisations and bodies that deliver and award qualifications. There will also be specific regulations which relate to your specialist subject and you will need to find out what these are. Examples include:

Control of Substances Hazardous to Health (COSHH) Regulations 2002 applies if you work with hazardous materials.

Food Hygiene Regulations 2006 applies to aspects of farming, manufacturing, distributing and retailing food.

Health and Safety (Display screen equipment) Regulations 1992 applies when using display screen equipment, for example, computers.

Manual Handling Operations Regulations 1992 covers anyone who transports or supports a load, including the lifting, putting down, pushing, pulling, carrying or moving by hand or bodily force.

Privacy and Electronic Communications (EC Directive) Regulations 2003 applies to all electronic communications such as e-mail and mobile phone messages.

Regulatory Reform (Fire Safety) Order 2005 places the responsibility on individuals within an organisation to carry out risk assessments to identify, manage and reduce the risk of fire.

Reporting of Injuries, Diseases and Dangerous Occurrences (RIDDOR) Regulations 1995 requires specified workplace incidents to be reported.

Codes of Practice

Codes of practice are usually produced by organisations, associations and professional bodies. They can be mandatory or voluntary and you will need to find out which are applicable to your job role. If you belong to any professional associations, they will usually have a code of practice for you to follow. For example, the Society for Education and Training (SET) in England (which is part of the Education and Training Foundation) has a Code of Practice. There are other professional associations such as the Chartered Institute for Educational Assessors (CIEA), the Institute for Leadership and Management (ILM) and the Institute of Training and Occupational Learning (ITOL), which you could join.

Your organisation should have documented codes of practice which you will need to find out about and follow, such as (in alphabetical order):

- acceptable use of information and communication technology (ICT)
- code of conduct
- confidentiality of information
- conflict of interest
- disciplinary
- duty of care to learners, including personal development, behaviour and welfare
- duty to prevent radicalisation
- environmental awareness
- lone working
- management of information and records
- misconduct
- sustainability.

Policies and procedures

There will be several policies and procedures in your organisation with which you should become familiar. It might be that your job role will require you to create or update a policy and/or a procedure at some point, for example:

- access and fair assessment

- appeals and complaints

- confidentiality of information

- copyright and data protection

- equality and diversity

- health, safety and welfare (including Safeguarding and Prevent Duty)

- internal quality assurance

- plagiarism and cheating

- malpractice.

Think of the policy as a statement of intent and the procedure as how the policy will be put into action. Policies don't need to be long or complicated, but should provide a set of principles to help with decision making. Procedures should state who will do what and when, and what documentation or checklists should be used. Policies and procedures should help guide your job role and should reflect the vision and mission of your organisation. Please see Chapter 5 for information regarding an organisation's vision and mission.

Extension activity

Find out what legislation, regulatory requirements, codes of practice, policies and procedures relate to your role as an internal quality assurer for your particular subject. Which are the most important and why? How will they impact upon your role and the internal quality assurance process?

Concepts and principles of internal quality assurance

Concepts relate to ideas, whereas principles are how the ideas are put into practice. For the purpose of this chapter, they have been separated for clarity, however some concepts could also be classed as principles depending upon your interpretation.

Key concepts

Concepts are the aspects involved throughout the IQA process. They include the following (which are then explained in detail):

- accountability

- achievement

- assessment strategies

- confidentiality

- risk factors

- evaluation

- interim and summative sampling

- transparency.

You need to be *accountable* to your organisation to ensure you are carrying out your role correctly. Your assessors should know why they are being monitored and why their decisions are being sampled. You will also be accountable to the awarding organisation if you assess an accredited qualification or endorsed programme. For example, if you or an assessor did something wrong, the credibility of your reputation could be questioned.

You may be required to analyse *achievement* data and compare this to national or organisational targets. Any funding your organisation receives might also be related to achievements. It's always a useful evaluation method to keep a record of how many learners your assessors have and how many successfully achieve. If one assessor has a high number of learners who are not achieving, is it because of the assessor or some other factor?

Following the *assessment strategy* from the awarding organisation for your subject will ensure you are carrying out your role correctly, are holding or working towards the required qualifications, as well as supporting your assessors. The assessment strategy states what qualifications and experience trainers, assessors and IQAs must hold. You might also need to recruit and interview new assessors and ensure they meet the assessment strategy (if applicable). It would also be useful to have a succession plan in place in case an assessor leaves at short notice. You can't have learners without an assessor, and you may need to inform the awarding organisation of any changes in staff.

Confidentiality will ensure you maintain records in accordance with organisational and statutory guidelines such as the Data Protection Act 1998 and the Freedom of Information Act 2000.

There are many *risk factors* to take into consideration when planning IQA activities. For example, how experienced and qualified the staff are, whether assessment methods are appropriate, and whether all staff are interpreting the requirements in the same way. Risk management is covered in more detail as you progress through this chapter.

Evaluation of the assessment and IQA process should always take place to inform current and future practice. All aspects of the IQA cycle should be evaluated on a continuous basis.

Sampling should take place on an ongoing basis and not be left until the end of the programme. It should be *interim*, i.e. part way through, and *summative*, i.e. at the completion stage. If a problem is identified at the interim stage, there is a chance to put it right. The summative stage can check the full assessment process has been successfully completed and that all documents are signed and dated correctly.

To assist *transparency* you need to ensure that everyone who is involved in the assessment and IQA process clearly understands what is expected and can see there is nothing untoward taking place. That includes your own interpretation and understanding of what is being assessed, as well as that of your staff. There should be no ambiguity, i.e. everyone should know what is expected of them. Transparency is also about having nothing to hide and being open to scrutiny.

Auditable records must always be safely maintained throughout the IQA process for a set period of time. This is often three years, however you will need to find out what the requirements are at your organisation. The records can be manual and/or electronic, providing they meet data protection and confidentiality requirements. If your organisation is claiming funding, full and accurate records must be maintained to show what was claimed, when and why.

Key principles

Principles are based upon the concepts and relate to *how* the IQA process is put into practice.

One important principle is known by the acronym VACSR and should be followed when carrying out internal quality assurance activities.

- **V**alid – assessor decisions and feedback are relevant to what has been assessed.
- **A**uthentic – the work has been produced solely by the learner, and the assessor has confirmed this.
- **C**urrent – the learner's work is relevant at the time of assessment, as are the assessment records which have been used.
- **S**ufficient – the learner's work covers all of the requirements, and the assessor records are complete, legible and accurate.
- **R**eliable – the assessment process is consistent across all learners, over time and at the required level.

If the above are not checked, you will not be supporting your assessors correctly. This could lead them to think their practice is acceptable when in reality it might not be.

Key principles of internal quality assurance include (in alphabetical order):

- assessor competence – ensuring assessors are experienced and competent at their role, meeting the requirements of the assessment strategy (if applicable) and are maintaining their CPD
- communication – ensuring this takes place regularly with learners, trainers, assessors, other internal quality assurers, employers, witnesses and anyone else involved
- CPD – maintaining your own currency of knowledge and performance to ensure your practice is up to date
- equality and diversity – ensuring all assessment activities embrace equality, inclusivity and diversity, represent all aspects of society and meet the requirements of the Equality Act 2010
- ethics – ensuring the assessment and IQA process is honest and moral, and takes into account confidentiality and integrity
- fairness – ensuring assessment and IQA activities are fit for purpose, and planning, decisions and feedback are justifiable

- health and safety – ensuring these are taken into account throughout the full assessment and IQA process, carrying out risk assessments as necessary and following the Prevent Duty guidelines

- motivation – encouraging and supporting trainers and assessors to reach their maximum potential

- record keeping – ensuring accurate records are maintained throughout the teaching, learning, assessment and IQA processes

- SMART – ensuring all training and assessment activities are specific, measurable, achievable, relevant and time bound

- standardisation – ensuring the requirements are interpreted accurately and that all assessors and internal quality assurers are making comparable and consistent decisions

- strategies – ensuring a written IQA strategy is in place which clearly explains the full process of what will be internally quality assured, when and how. The strategy should be based on an IQA rationale, which will be explained after the following activity.

Activity

Look at the previous bulleted lists of concepts and principles, and list those which you feel are relevant to your role as an IQA. Explain what will be involved in each one that you have listed.

Following the key concepts and principles of internal quality assurance will ensure you are performing your role according to all relevant regulations and requirements.

The IQA rationale

A good IQA system will start with a written rationale: this is the reason *why* IQA takes place. A rationale will help maintain the credibility of what is being assessed as well as the reputation of your organisation.

For example, for your own rationale you might think it is good practice for all assessors and IQAs to hold a recognised qualification (whether the assessment strategy for the subject requires it or not).

Example rationale

All internal quality assurance activities will comply with internal and external organisations' requirements to assure the quality of assessment for all learners. All assessment decisions will be carried out by qualified assessors in each subject area and sampled by qualified internal quality assurers. This will ensure the safety, fairness, validity and reliability of assessment methods and decisions. It will also uphold the credibility of the qualification and reputation of the organisation.

Having a rationale will help ensure all assessment and IQA activities are robust, and that they are safe, valid, fair and reliable.

- Robust – the activities are strong and will endure the test of time.

- Safe – the activities used are ethical, there is little chance of plagiarism by learners, the work can be confirmed as authentic, confidentiality is taken into account, learning and assessment is not compromised, nor is the learner's experience or potential to achieve (safe in this context does not relate to health and safety but to the assessment methods used).

- Valid – the activities used are based on the requirements of the programme, qualification, work tasks, job description or role specification.

- Fair – the activities used are appropriate to all learners at the required level, taking into account any particular learner needs. Activities are fit for purpose, and planning, decisions and feedback are justifiable and equitable.

- Reliable – the activities used will lead to a similar outcome with similar learners.

An IQA strategy should then be produced based on the rationale. The strategy will specifically state which monitoring activities will take place. This is covered in more detail in Chapter 2.

The IQA process

If there is no external formal examination taken by learners, there has to be a system of monitoring the performance of trainers and assessors, and the experiences of the learners. If not, assessors might make incorrect judgements or pass a learner who hasn't met the requirements, perhaps because they were biased towards them or had made a mistake. Assessment and IQA systems should be monitored and evaluated continuously to identify any actions for improvement, which should then be implemented. This also includes the CPD of assessors and internal quality assurers.

An internal quality assurer should be appointed to carry out the quality role within an organisation where there are assessment activities taking place.

As a minimum, the internal quality assurer should:

- plan what will be monitored, from whom and when

- observe trainer and assessor performance and provide developmental feedback

- sample assessment records, learners' work and assessor decisions

- meet with learners and others, for example, witnesses from the workplace

- facilitate the standardisation of assessor practice

- support assessors and link their development needs to CPD.

If there is more than one internal quality assurer for a particular subject area, one person should take the lead role and co-ordinate the others. They are often referred to as a lead IQA. Please see Chapter 5 for further details of this role. All internal quality assurers should standardise their practice with each other to ensure they interpret the requirements in the same way.

If IQA does not take place, there are risks to the accuracy, consistency and fairness of assessment practice. This could lead to incorrect decisions and ultimately disadvantage the learners.

Extension activity

Look at the previous bullet list. Are there any other aspects you feel you should carry out as an IQA? If there are other IQAs for your subject area, ask them what they do and why. Find out who the lead IQA is, talk to them and make sure you are both aware of who will do what.

Risk management

Whenever possible, it's best to be proactive by managing any risks that might occur throughout the assessment process, rather than being reactive to a situation after the event. You really need to be in regular contact with your assessors to build up a trusting relationship, whereby they feel able to inform you of any concerns they might have. If you are not aware of any situations which might pose a risk to assessment, a situation could become quite serious. It's important to monitor and manage risks to ensure adequate support can be given to assessors and that learners are not disadvantaged in any way. Just ask yourself what could possibly go wrong, and if you can think of something, then there is a risk.

There are many risk factors to take into consideration when planning the IQA activities you will carry out. These risks relate to disadvantaging the learner in some way through the assessment process, rather than health and safety risks. Being aware of these might enable you to prevent some situations occurring. If you know they are likely to occur, you will need to carry out further IQA monitoring activities to check for them. These will be explained in Chapter 2.

Possible risks (in alphabetical order):

- a lack of confidence by the assessor to make correct decisions

- a lack of standardisation activities leading to one assessor giving more of an advantage to a learner than another assessor of the same subject

- a learner's lack of confidence or resistance to be assessed

- action points from external reports not being carried out by the target date

- an assessor not taking into account a learner's particular needs

- an unsuitable environment for assessment to take place

- answers to questions being obtained inappropriately by learners which leads to cheating

- assessing written work too quickly and not noticing errors, plagiarism or cheating

- assessor expertise, knowledge and competence not up to date or whether new staff are working towards an assessor qualification. Unqualified staff might need their decisions countersigning. Staff should have appropriate job descriptions and development plans otherwise they won't know what they are doing

- assessors using leading questions to obtain the correct answers they require

- assistive technology for learners with particular needs is used wrongly or used to give too much support

- awarding organisations prescribing assessment methods which might not complement the qualification, a learner's needs or the learning environment

- changes to qualifications, standards, documents, records, policies and procedures: staff need to be kept up to date

- feedback to the learner which is unhelpful or ineffective

- high turnover of staff resulting in inconsistent support to learners

- how quickly (or how slowly) learners complete with a particular assessor

- instructions too complex or too easy for a learner's ability

- insufficient or incorrect action/assessment planning

- lack of resources or time to perform the assessment role correctly

- learners creating a portfolio of evidence which is based on quantity rather than quality, i.e. submitting too much evidence which does not meet the requirements

- locations of learners and assessors which might make them inaccessible at times

- marking and grading carried out incorrectly by assessors

- misinterpreting the assessment requirements and/or criteria (by learners and assessors)

- numbers of learners allocated to assessors is too high, leading to rushed assessments

- resources and equipment accessibility, availability and safe use

- time pressures and targets put upon assessors and learners

- type of qualification or programme being assessed, problem areas or units

- types of evidence provided by learners e.g. a reliance on too many personal statements which are not backed up by assessment decisions

- unreliable witness testimonies from the workplace, lack of support to witnesses

- unsuitable assessment methods i.e. an observation when questions would suffice

- unsuitable assessment types i.e. summative being used instead of formative

- unwelcome disruptions and interruptions when assessing, such as noise or telephone calls

- use of technology and how reliable it is for assessment purposes

- whether bilingual assessments could lead to any issues or language barriers impeding communication

- whether evidence and records are stored manually or electronically in case they can be accessed by unsuitable people

- whether the learners have been registered with an awarding organisation (if applicable) as assessment decisions might not be valid otherwise.

All the above can impact upon the amount of IQA activities which need to be carried out, with whom and when. You should also consider any risks regarding your own role i.e. making an invalid decision or giving inappropriate feedback because you are not up to date with what is being assessed. You need to make sure your own practice is current and that your judgements are valid and reliable.

Malpractice

Malpractice is about someone doing something wrong which could be classed as professional negligence, for example, an assessor signing off units which are incomplete, or doing some work for their learners. If you are quality assuring an accredited qualification or endorsed programme on behalf of an awarding organisation, they will have guidance for you to follow.

Activity

If you are currently working as an IQA, obtain a copy of the awarding organisation's quality manual or qualification specification for your subject. Find out what their advice is regarding malpractice and check this with your own centre's policy and procedure.

Malpractice could be intentional or accidental and you would tactfully need to find out which if you suspect it is taking place. Having some knowledge of what might occur will help you watch out for it.

Areas where malpractice could occur include (in alphabetical order):

- an internal quality assurer overruling an assessor's decision (due to pressures to meet targets) when the assessor did not pass the learner

- assessment records being signed off when assessments did not take place

- certificates being claimed for learners who did not exist or who have not yet completed

- dates of commencement and achievement not agreeing with those that the learners tell you

- dates on records not matching those when the activities took place

- learners' work and supporting records which you have requested are not available or belong to someone else

- minutes of meetings being produced when they didn't actually take place

- signatures on documents not matching those of the people concerned

- standardisation records being completed for activities that did not take place.

Plagiarism and authenticity

When internally quality assuring the assessed work of learners, you need to be aware that some learners, intentionally or not, might plagiarise the work of others or copy from the internet without acknowledging the source. Plagiarism is the wrongful use of someone else's work. Authenticity is the rightful and confirmed use of your own work. You need to be aware of learners colluding or plagiarising work, particularly now that so much information is available via the internet. Assessors should make sure that learners take responsibility for referencing any sources in all work submitted, and that they may be required to sign a declaration or an authenticity statement. If you suspect plagiarism, you could type a few of the learner's words into an internet search engine or specialist software and see what appears. You would then have to talk to your assessor to find out if their learner had done it intentionally or not.

Example

Sara, the IQA, was sampling the work of her assessors. She noticed one learner's assignment contained several extracts from the internet which had not been acknowledged. She knew this as she had recently been researching the topic online. Sara approached Peter, the assessor, to ask why he had not commented on it. Peter confessed he hadn't noticed it as he had only skim read it. Sara informed Peter of the importance of thorough assessment and gave him a copy of the plagiarism policy. She asked him to talk to his learner and let her know the outcome. Sara then added 'plagiarism' to the agenda for discussion at the next team meeting.

Unfortunately, some learners do cheat, copy or plagiarise the work of others. Sometimes this is deliberate, and at other times it is due to a lack of knowledge of exactly what was required, or a misunderstanding when referencing quotes within work. You need to ensure your assessors are aware of this. If an incorrect assessment decision is given, and this is not noticed, a learner might achieve a certificate for a qualification that they have not legitimately accomplished.

Some ways which you and your assessors can check the authenticity of learners' work include:

- spelling, grammar and punctuation – if you know the learner speaks in a certain way at a certain level, yet their written work does not reflect this

- work that includes quotes which have not been referenced – without a reference source, this is direct plagiarism and could be a breach of copyright

- word-processed work that contains different fonts and sizes of text – this shows it could have been copied from the internet or from someone else's electronic file

- handwritten work that looks different to a learner's normal handwriting or is not the same style or language as normally used, or work which has been word-processed when they would normally write by hand

- work that refers to information which has not been taught or is not relevant.

If assessors and learners are completing and/or submitting documents electronically, there might not be an opportunity to add a real signature to confirm authenticity. However, many companies are now accepting a scanned signature, electronic signature or an e-mail address instead. This might be acceptable providing the identity of the person has been confirmed and a record kept of the original signature.

The Copyright, Designs and Patents Act 1988 is the current UK copyright law. Copying the work of others without their permission would infringe the Act. Copyright is where an individual or organisation creates something as an original and has the right to control the ways in which their work may be used by others. Normally the person who created the work will own the exclusive rights. However, if the work is produced as part of your employment, for example, if you produced a guide for your assessors, then normally the work will belong to your organisation. Learners may be in breach of this Act if they plagiarise or copy the work of others without making reference to the original author.

Learner risks

As an IQA, you need to support your assessors to minimise any risks to learners, for example, assessors putting unnecessary stress upon learners, over-assessing, under-assessing, or being unfair and expecting too much too soon. Some learners might not be ready to be observed for a practical skill, or feel so pressured by target dates for a theory test that they resort to colluding or plagiarising work. If learners are under pressure or have any issues or concerns that have not been addressed, they could be disadvantaged and might decide not to continue with the programme.

Assessor risks

Some assessors might feel pressured to pass learners quickly due to funding and targets. Others might unknowingly offer favouritism or bias towards some learners over others. A personal risk to assessors could be if they carry out assessments in the work environment and visit places with which they are not familiar. They might need to travel early or late in the dark, find locations on foot, take public transport or drive to areas they are not familiar with. There should be a lone worker policy at your organisation for their protection and you will need to make sure they are aware of it.

The type of employment contract assessors have might also pose a risk, for example, if part time, working for more than one organisation, or working for an agency or on a free-lance basis. This could cause some confusion; therefore you will need to ensure all your assessors are familiar with the procedures at your organisation. Standardisation of practice might also be difficult if assessors are not all in the same location or working for the same organisation. There might also be the risk of assessors feeling pressured to pass learners if they are paid by the number of learners who complete the qualification.

Other risks

Assessors who visit learners in the work environment might come across employers who are not supportive and may put barriers in their way. For example, someone might make it difficult for an assessor to visit at a certain time to carry out a formal assessment. Careful planning and communication with everyone concerned will be necessary.

You will need to be aware if assessors are assessing close friends or relatives. You might want to sample more of their work, or another impartial assessor could countersign their decisions. If it is an accredited qualification, the awarding organisation will be able to give you guidance regarding this. They might consider it a conflict of interest and not allow it to take place.

Extension activity

What risks do you feel you will encounter as part of your IQA role? Choose four risks and consider how you could prevent them occurring.

Maintaining and improving the quality of assessment

If you have a robust quality system, you will maintain the credibility of what is being assessed, as well as the reputation of your organisation. This will give learners confidence that they will be assessed correctly. Carrying out observations of your assessors, sampling their learners' work, their assessment decisions and records, will all help ensure IQA is an effective, valid and reliable process. However, it needs to remain effective and cannot be allowed to decline in any way. You therefore need to ensure that you monitor and improve assessment and IQA processes and systems on an ongoing basis.

You must maintain an audit trail of everything you do; creating a sampling plan of what you are going to do will be the starting point. If you plan to IQA something in March and don't sample it until April, that's absolutely fine. However, the actual date you carried out the activity must be added to the plan. Having *March* on the sample plan but *12 April* as the actual sampled date is reflecting reality. Not everything will occur when you planned for it, perhaps due to holidays, the late submission of learner work, absence or illness. If you do change any dates, don't be tempted to alter them or use correction fluid. Just cross out one date and write in another so that the original can still be seen. If you save your plans electronically, resave the changes as a different version so that you can access the previous versions if required. You might need to justify your reasons to auditors as to why there have been any changes. Sample planning will be covered in detail in Chapter 2.

To help improve the quality of assessment you need to ensure your assessors are maintaining their occupational competence for the subject area they are assessing. You also need to make sure they are fully competent with your organisation's systems, policies and procedures, as well as assessment practice in general. If your assessors are qualified, they should be operating at the required standard to support their learners. If you have assessors that are currently working towards an assessor qualification, their decisions may need to be countersigned by another qualified assessor in the same subject area. As the internal quality assurer, you cannot countersign any assessment decisions which you will then internally quality assure, as this is a conflict of interest. However, you will need to ensure any countersigning by other assessors has correctly taken place when sampling the work of unqualified assessors. Your awarding organisation should be able to give you further guidance on this if necessary, as not all qualifications require countersigning to take place.

When a new trainer or assessor commences at your organisation, it might be your responsibility to induct them to ensure they understand their job role and the requirements of what will be assessed. The following checklist might prove useful with new staff, and with existing staff from time to time to refresh their practice.

IQA checklist for new staff

☐   *Tour of the organisation and training/assessment locations*

☐   *Introductions to other relevant staff*

☐   *Organisation's policies and procedures, vision and mission statement*

☐   *Curriculum vitae, CPD and qualifications checked*

☐   *Sample signature taken and photo identification checked*

☐   *Explanation of the qualification standards or criteria to be used*

☐   *Details of learners they will be responsible for*

☐   *Assessment strategy and assessment documentation: how to access and complete*

(Continued)

- ☐ *Internal and external quality assurance procedures*

- ☐ *Target date agreed for achieving relevant qualifications (if applicable)*

- ☐ *Mentor and/or countersignatory identified if applicable*

- ☐ *Other aspects such as access to resources, photocopying, administrative support, travel expense claims, pay claims*

You should also discuss the training and development needs of your staff to ensure they are keeping up to date with their knowledge and practice. This is relevant not only for their subject, but also training and assessment practice in general. Discussing this with them will help identify any requirements they have, which can then be met through CPD. Improving your assessors' performance should lead to an improvement in the quality of assessment.

Activity

Look at the previous checklist. What aspects do you feel you would carry out with a new assessor and why? Consider other activities which you might do with an existing assessor and state why.

As an IQA, it's a good idea to regularly review the documents, records, templates and checklists used by assessors and IQAs. It might be that two forms are currently being used which could be merged into one form, or some questions on a document are no longer relevant. A standardisation activity between staff could be arranged to review and update all the documentation and records used. It's useful to include the date or a version number on each document as a footer and keep a separate list of these details for tracking purposes. That way, you can quickly see if everyone is using the most up-to-date version.

Example

Magda sampled the assessment documents used for unit 104 from each of her four assessors. She found one of the assessors had used completely different documentation to the other three. Magda random sampled further units from this assessor and found the same. She was then able to give feedback to the assessor as to where the current versions could be located and how they should be used.

Encouraging staff to access documents electronically from a central system when they need them, rather than keeping a stack of hard copies, will help ensure they are using the most up-to-date version as well as being environmentally aware.

Extension activity

How could you check that the assessors you will be responsible for are all up to date regarding the subject being assessed and assessment practice in general? What would you recommend they do to ensure they keep up to date?

The role of technology in internal quality assurance

Information and communication technology (ICT) can be used to support and enhance the IQA process. This is particularly useful when the internal quality assurer is located in a different area from their assessors. Learners and assessors could upload completed work and records to a virtual learning environment (VLE), electronic portfolio (e-portfolio) or cloud-based system. This would enable the internal quality assurer to sample various aspects remotely at a time to suit. Reports could then be completed electronically, uploaded to the system or e-mailed as required.

Communication through e-mail or web-based forums can simplify the contact process between the internal quality assurer and their assessors. There will be times when people are not available at the same time, either for a meeting or a telephone call. Using ICT enables messages to be left which can be responded to when convenient. However, this method of communication does rely on people accessing their messages on a regular basis.

Meetings and standardisation activities could also take place remotely, for example, through video/teleconferencing or webinars. Everyone does not need to be in the same room at the same time for activities to be effective. Materials could be produced and circulated electronically prior to the remote meeting and then discussed when everyone is accessible.

Never assume that staff are familiar with how to use the various aspects of ICT. Training sessions may need to be carried out and resources might need to be updated to support their use.

Activity

Consider what aspects of the IQA role could be carried out using technology. Make a list of the equipment, devices, programs and mobile applications which would be needed. How could you make sure everyone had access to them and knew how to use them? It could be that you would expect your staff to use their own devices if resources are not available.

Some examples of using ICT for internal quality assurance activities, with the permission of the people involved, include (in alphabetical order):

- making digital recordings or videos of role-play activities or case studies. For example, a film of an assessor making an assessment decision and giving developmental feedback to a learner. Assessors could view them remotely to comment on strengths and limitations

- making visual recordings of how to complete documents and reports. If an assessor is unsure how to complete a document they could access a video to see an example

- observing live assessment activities via an online visual communication program

- recording verbal information, making podcasts or visual recordings of conversations, meetings and/or information regarding updates and changes, to be viewed later by team members

- using a mobile phone/smartphone, tablet or digital camera to record an assessor activity. This is useful if the internal quality assurer cannot be present at the time – the assessor could make a recording and forward it on to them

- using cloud-based or online systems to store, access and revise various documents

- using e-mails with integrated video facilities to send visual messages

- using web conferencing to talk to assessors, learners and witnesses if they are quite a distance away

- using webinars to view presentations or review software packages with team members, enabling them to remain in their own locations rather than travel to a central location.

Example

Ling, the IQA, regularly sends e-mails to communicate with her assessors. As she has a smartphone with a camera, she thought she would try something different and use a video e-mail. She accessed a free program at www.mailvu.com. This enabled her to make a quick recording and e-mail it to her assessors. They said her visual message had more impact than just reading text.

Digital literacy is the term often used for using and benefiting from information and communication technology. ICT can be a useful tool for learning, assessment and quality assurance activities. Your assessors will need to know how technology can assist when differentiating assessment activities to meet a particular learner's requirements, for example, using a screen reader or text enlargement software for a learner who is partially sighted. *Assistive or adaptive technology* is the term used to denote devices and their use for people with disabilities or difficulties. It can lead to greater independence by providing enhancements to, or changing the methods of, use. This should enable learners to accomplish tasks they might not have been able to do without it. You might therefore need to find out if any of your assessors need relevant training to support their learners.

Alternatively, your assessors might wish to use assistive or adaptive technology for their own role.

There should be a code of practice for the use of ICT for both staff and learners to follow. You will need to locate and read this and ensure your assessors are familiar with it too. For example, if assessors are using online applications for assessment purposes, they should not be accessing their personal social media accounts at the same time. Following the code will help ensure that the use of technology is reasonable and safe.

Examples of technology for learners, assessors and IQAs (in alphabetical order):

- blogs, chat rooms, social networking sites, webinars and discussion forums to help assessors communicate with each other
- BYOD (bring your own device) – learners and assessors use their own devices
- cloud storage facilities which assessors can use to upload, access and update materials from various devices connected to the internet
- computer facilities for assessors to create resources, save and back up their work
- digital media for visual/audio recording and playback
- electronic portfolios for learners, assessors and IQAs to access
- e-mail for communication, either text or visual
- internet access for researching various topics
- mobile phones and tablets for taking pictures, video and audio clips, and communicating with others
- networked systems to allow access to applications and documents from any computer linked to the system
- online discussion forums which allow asynchronous (taking place at different times) and synchronous (taking place at the same time) discussions
- scanners for copying and transferring documents to a computer
- VLEs to access and upload materials, assessment activities and feedback
- web cameras or video conferencing if you can't be in the same place as your assessors and you need to see and talk to them.

Extension activity

Challenge yourself to use a piece of equipment or a new program or application which you have not used before. For example, you might like to create a VLE for learners and assessors to use. There are a few free ones available such as Simple VLE. It can be accessed at: www.simplevle.com/v3/index.jsp

Summary

Following the key concepts and principles of internal quality assurance will ensure you are performing your role as an IQA according to all the relevant requirements, regulations and standards. It will also help you to support your trainers and assessors, and ensure all learners receive a high quality service.

You might like to carry out further research by accessing the books and websites listed at the end of this chapter.

This chapter has covered the following topics:

- Quality assurance

- Roles and responsibilities of an internal quality assurer

- Concepts and principles of internal quality assurance

- Risk management

- Maintaining and improving the quality of assessment

- The role of technology in internal quality assurance

References and further information

Gravells, A. (2016) *Principles and Practices of Assessment* (3rd edition). London: Learning Matters SAGE.

JISC (2010) *Effective Assessment in a Digital Age: A Guide to Technology-enhanced Assessment and Feedback.* Bristol: JISC Innovation Group. Available at: www. jisc.ac.uk/digiassess

Pontin, K. (2012) *Practical Guide to Quality Assurance.* London: City & Guilds.

Read, H. (2012) *The Best Quality Assurer's Guide.* Bideford: Read On Publications Ltd.

Sallis, E. (2002) *Total Quality Management in Education* (3rd edition). Abingdon: Routledge.

White, J. (2015) *Digital Literacy Skills for FE Teachers.* London: SAGE.

Wilson, L. (2012) *Practical Teaching: A Guide to Assessment and Quality Assurance.* Andover: Cengage Learning.

Wood, J. and Dickinson, J. (2011) *Quality Assurance and Evaluation in the Lifelong Learning Sector.* Exeter: Learning Matters.

Websites

Assistive technology – www.washington.edu/doit/resources/popular-resource-collections/accessible-technology

Council for the Curriculum, Examinations and Assessment in Northern Ireland (CCEA) – http://ccea.org.uk

Chartered Institute for Educational Assessors – www.ciea.org.uk

The Copyright, Designs and Patents Act 1988 – www.legislation.gov.uk/ukpga/1988/48/contents

Counter-Terrorism and Security Act 2015 –www.legislation.gov.uk/ukpga/2015/6/contents/enacted

Data Protection Act 1988 – www.legislation.gov.uk/ukpga/1998/29/contents

The Department for Children, Education, Lifelong Learning and Skills (*DCELLS*) in Wales – http://gov. wales/topics/educationandskills/?lang=en

Disability and the Equality Act 2010 – http://tinyurl.com/2vzd5j

Disclosure and Barring Service – www.gov.uk/government/organisations/disclosure-and-barring- service/about

Dropbox online file sharing – www.dropbox.com

Education and Training Foundation – www.et-foundation.co.uk

Equality and Human Rights Commission – www.equalityhumanrights.com

FELTAG Report (2014) *Paths forward to a digital future for Further Education and Skills* http://feltag. org.uk/

Freedom of Information Act 2000 – www.legislation.gov.uk/ukpga/2000/36/contents

Health and Safety Executive – www.hse.gov.uk

Health and Safety resources – www.hse.gov.uk/services/education/information.htm

Initial assessments for ICT – www.tes.co.uk/teaching-resource/ict-initial-assessments-6177727

Joint Information Systems Committee (JISC) – www.jisc.ac.uk

Ofqual – www.ofqual.gov.uk

Ofsted – www.ofsted.gov.uk

Online free courses in various subjects – www.vision2learn.net

Online games – www.npted.org/schools/sandfieldsComp/games/Pages/Game-Downloads.aspx

Online presentations – www.prezi.com

Open Office computer software – www.openoffice.org/download/

Pinterest – https://uk.pinterest.com

Plagiarism – www.plagiarism.org and www.plagiarismadvice.org

Prevent Duty and Safeguarding resources – www.preventforfeandtraining.org.uk

Puzzle software – www.about.com

        www.crossword-compiler.com

        http://hotpot.uvic.ca

        www.mathsnet.net

Scottish Qualifications Authority (SQA) – www.sqa.org.uk

Self-evaluation – www2.warwick.ac.uk/services/ldc/resource/evaluation/tools/self/

Skype online visual communication site – www.skype.com

Social media – www.facebook.com

        https://instagram.com

        www.linkedin.com

        https://plus.google.com

        www.twitter.com

Society for Education and Training (SET) – https://set.et-foundation.co.uk

Surveys and questionnaires – www.surveymonkey.com

Using computers and technology – http://digitalunite.com/

Using Microsoft programs – www.reading.ac.uk/internal/its/training/its-training-index.aspx

Video e-mail – www.mailvu.com

VLE – www.simplevle.com/v3/index.jsp

YouTube – www.youtube.com

2   PRACTICES OF INTERNAL QUALITY ASSURANCE

Introduction

Carrying out internal quality assurance activities will help ensure assessment decisions are accurate, consistent, valid and reliable. The tasks you carry out should always be performed in an organised and professional manner, with full records being maintained.

Internal quality assurance might just be a small part of your job role; you might also be a trainer and assessor, and/or carry out other relevant duties according to your job description.

This chapter will explore how you can plan for and carry out various internal quality assurance activities, including preparing for an external quality assurance visit.

This chapter will cover the following topics:

- Internal quality assurance planning
- Internal quality assurance activities
- Making decisions
- Providing feedback to assessors
- Record keeping
- Evaluating practice

Internal quality assurance planning

When you are appointed as an internal quality assurer (IQA), you should be given a job description which will help you understand the requirements of your role. If you don't have one, then working towards the requirements of an appropriate internal quality assurance (IQA) qualification or following national occupational standards will ensure you are performing your role effectively. You should find out if there are any other IQAs in the same subject area as yourself within the same organisation. If so, you can communicate with them to find out how you go about planning what you will do, and to ask for any advice and support.

You will need to familiarise yourself with your organisation's internal quality assurance policy, rationale and strategy. The policy and rationale have been covered in Chapter 1; the strategy will be covered later in this chapter. These documents might already be available;

if not, you might have to produce your own. You will also need a copy of what is being assessed i.e. the syllabus, qualification handbook or role specification if you don't already have a copy. You will need to refer to this when planning and carrying out your IQA activities. You will also need to access the IQA documentation which you will use. If none is available, various examples are given throughout this chapter which you could use as a starting point.

Example

Vince has been employed as a motor vehicle maintenance assessor for five years. He has recently passed the internal quality assurance qualification and has been promoted to an IQA role due to the retirement of the previous IQA. This will enable him to support the five assessors in his team and monitor their practice. He has worked as part of the team therefore he knows the assessors and is familiar with the qualification requirements. His first task will be to familiarise himself with the IQA policy, rationale, strategy and documentation. He can then start to plan the monitoring activities he will carry out.

A large part of your role is about monitoring assessor practice and you will need to plan what you will do and when. It's therefore advisable to be fully familiar with assessment approaches for the subject you will internally quality assure. This will help you support your assessors and know what you are looking for when sampling. Before commencing the planning process, you could use a checklist like the one here to help you prepare.

Checklist – preparing for your role

☐ *Do all assessors have an up-to-date copy of what they are assessing their learners towards? Do you have a copy?*

☐ *Are there any areas which might cause concern, i.e. aspects which are difficult for learners to achieve, or involve complex activities?*

☐ *Are all assessors qualified and/or experienced in their subject area? You may need to check that their curriculum vitaes (CVs) and certificates conform to any requirements, and that continuing professional development (CPD) is ongoing.*

☐ *Do any assessors need to take an assessor qualification or gain any further experience? If so, you may need to arrange training and development activities for them. You may also need to ensure unqualified assessors have their decisions countersigned by another qualified assessor in the same subject area (applicable).*

☐ *Are there adequate policies and procedures such as appeals and complaints? If not, you may need to produce or update them, and ensure assessors and learners are familiar with them.*

☐ *Is there a suitable IQA policy, rationale and strategy? If not, you will need to create them or update the previous ones.*

☐ *Have learners been allocated to assessors? If not, you need to do this fairly according to each assessor's location and workload.*

☐ *Are the assessment and IQA records and documents suitable or do they need updating? If there aren't any, you will need to create them.*

Activity

Make a list of any other aspects you need to check to help you prepare for your role, in addition to those in the previous checklist. Find out who can give you any support and advice should you need it. If you have not already done so, obtain a copy of what is being assessed i.e. the syllabus, qualification handbook or role specification, along with copies of the IQA documentation you will use.

If you are internally quality assuring an accredited qualification or endorsed programme, it will be externally quality assured. This means a member of the relevant awarding organisation (AO) will visit to ensure you are compliant with their requirements. This person will be known as an external quality assurer (EQA). Some other terms for this role are used by different AOs, for example, *external quality consultant* or *standards' verifier*. You will need to find out who this person is and make contact with them if it's part of your role to do so. Alternatively, it might be another person's role such as an administrator or exam officer, and you will need to communicate via them.

The amount and frequency of EQA monitoring activities will depend upon how active your organisation is with their learners. If the EQA has confidence in your systems and the way your organisation operates, they might not carry out a visit but conduct a remote sample instead. This means you will post learners' work, assessment and IQA records to the EQA, or give them remote access to electronic portfolios and records. You should share all the feedback you receive from the EQA with your team, along with any action and improvement points. Preparing for an EQA visit will be covered at the end of the next section within this chapter. You can also find out about the EQA's role in Chapters 3 and 4.

Once you have all the information and documentation you need, you can begin to plan and monitor assessor practice. You should have the support of the management within your organisation and have sufficient time and resources to perform your role effectively. If not, this could jeopardise the quality assurance process and disadvantage the learners. If you are concerned that you do not have enough time to perform your role effectively, you will need to discuss this with your manager. Never feel pressured to rush aspects of your role due to a lack of time, as mistakes might be made. Meeting your assessors soon after being appointed will help you get to know them. You will also need to make sure your assessors have adequate time for their role too.

Activity

Make a list of the assessors you are responsible for and ensure you have their contact details and locations. If you are responsible for an accredited qualification or an endorsed programme, find out the EQA's name and contact details. Have a look at the last report from them to check that any action points are not outstanding. If any have not been completed, you will need to find out why and then see if they can be completed. If not, your organisation might be placed on a sanction by the AO.

To perform your IQA role fully, you need to create sample plans, monitor assessment activities and collect various types of information. You might have several sample plans for different activities, for example, a plan to observe your assessors, a plan to sample their assessed work and a plan for meetings and standardisation activities. You might be able to combine the plans if there are only a small number of assessors in your team. To help you plan effectively, you need to have an IQA strategy on which to base your activities.

IQA strategy

Think of the IQA strategy as the starting point for all the activities, monitoring and sampling which you will carry out. It should be a written statement of what will be carried out and is based on the IQA rationale and any identified risk factors (covered in Chapter 1). Having a strategy will help you plan what will be monitored and when, and ensure that your systems are fit for purpose. If you are quality assuring an accredited qualification, it will be a requirement of the awarding organisation that you have a written strategy.

Systems should be in place to ensure all the key IQA concepts and principles are met, and that monitoring and sampling are effective. The strategy might be produced by your organisation or it might be your responsibility to write it.

Example

IQA strategy for Level 2 Customer Service (one internal quality assurer, four assessors, 100 learners):

The IQA will:

- *observe each assessor every six months in different locations to cover different assessment methods (new assessors will be observed more as necessary)*

- *talk to a sample of learners and witnesses*

- *sample at least five assessed units from each assessor across a mix of learners (new assessors will have a higher sample rate)*

- *chair a bi-monthly team meeting*

- *facilitate regular standardisation activities to cover all units over a period of time*

- *maintain full records of all IQA activities*

- *liaise with the EQA and implement external quality assurance action points.*

The amount of detail you write in your strategy will be dependent upon the experience of your assessors and the types of activities you will carry out. When planning your strategy you should take into account factors such as:

- assessment locations: on or off the job, college, training room or other environment

- assessment methods: are they robust, safe, valid, fair and reliable; are they complex and varied; do they include online assessments; are witnesses used; does holistic assessment track achievements across different units?

- assessors: availability for observations and meetings (some assessors could be located at a distance or have other jobs – can activities take place remotely via the internet?); assessors' experience, qualifications, workload and caseload (experienced assessors could be sampled less than new assessors)

- learners: location, full time/part time, ethnic origin, gender, specific needs (do assessors need to adapt any assessment methods for any particular needs?)

- qualification or criteria to be assessed: are assessors familiar with these; have assessors standardised their interpretation of them; is the content due to be revised; is there a risk of plagiarism or malpractice?

- types of records to be completed (manual or electronic).

Extension activity

Write a strategy for the subject you will internally quality assure. Consider what you will need to do and how you will do it. If you are not currently quality assuring, create a hypothetical strategy based on the previous bullet points.

Internal quality assurance activities

Internal quality assurance activities consist of many different events which need to be planned for and then carried out, for example, observing trainer and assessor practice, holding team meetings, facilitating standardisation activities and preparing for an external quality assurance visit.

Once you have your IQA strategy, you will need to use your organisation's documents to create sampling plans for the activities you will carry out. The plans will show what you aim to do over a period of time. If sampling plans are not available you will need to design your own, perhaps based on those in this chapter.

As a minimum, you will need:

- an observation plan

- a meeting and standardisation plan

- a sample plan and tracking sheet.

You could have separate plans for each activity if there are a lot of assessors and learners, or combine them if numbers are small. Your plans should show continual activity over the period of assessment. If IQA is carried out towards or at the end of the assessment process there is little opportunity to rectify any concerns or issues. There are lots of variables to take into consideration when creating your plans, for example, the experience and location of your assessors, the number of learners they have, the complexity of what is being assessed and the use of witnesses. You need to consider all of this in your strategy before planning your activities.

Observation plan

You should observe each of your assessors over a period of time to ensure they are carrying out their role effectively, supporting their learners adequately, making correct decisions and giving developmental feedback. You may also need to observe training sessions if relevant. When planning to observe, you will need to take into consideration your staff's experience, the number of learners they have and the different locations where they train and assess. When you have built up confidence in their performance, you could carry out fewer observations. If you have inexperienced or newly qualified staff you might want to observe them more. If you plan well in advance and liaise with your staff as to what they will be doing and when, you can ensure you cover a variety of what takes place.

Observing the same unit across different staff will aid the standardisation process. You might find your staff do things quite differently and this way you will be able to discuss ways of ensuring consistency of practice. It is the ideal time after an observation to talk to learners (and witnesses if applicable), and you should maintain accurate records of these discussions. An observation plan, as in Table 2.1, helps everyone see when the activities will take place. However, as it's a plan, dates may change and that's fine. This example shows the month in which an observation will take place; however, the actual date will be agreed and added nearer the time. An observation report will be completed, and if learners and witnesses are interviewed afterwards, a further report should also be completed. Examples of these reports can be found as you progress through this chapter.

Table 2.1 shows a plan for observing four assessors, two of whom have already been observed, and the dates of the observations have been added to the plan. P. Jones and M. Singh are qualified, J. Smith is unqualified and S. Hans is newly qualified. The IQA has decided the latter two will be observed more often until he has confidence in their work. Unit 101 will be sampled from every assessor to ensure a standardised approach is taking place.

Table 2.1 Example observation plan

Observation plan

Qualification: Level 2 Customer Service

IQA: H. Rahl

Assessor	Jan	Feb	Mar	Apr	May	Jun	Jul	Aug	Sep	Oct	Nov	Dec
P. Jones qualified	Unit 101 31.01						Unit 102					
M. Singh qualified		Unit 101 18.02						Unit 103				
J. Smith unqualified			Unit 101		Unit 102			Unit 103	Unit 104			
S. Hans newly qualified				Unit 101				Unit 103		Unit 105		

For the purposes of future proofing this textbook, a year has not been added to dates in the tables, reports and checklists. You should always add the year as well as the day and month to any records you complete.

Meeting and standardisation plan

Regular team meetings between all assessors and internal quality assurers should take place to discuss general issues and learners' progress. A plan will reflect when these will take place and ensure staff are available to attend. An agenda should be issued in advance and minutes taken and distributed after the meeting. If someone cannot attend, they should be given access to a copy of the minutes and be given the opportunity to discuss the content with you at some point. Meetings could take place virtually via teleconferencing, or be visually recorded and then viewed later.

Part of the IQA process is to ensure assessors are standardising their practice and are being consistent. These activities can be timetabled to take place as part of the team meeting or as a separate event, and records must be maintained. Besides the standardisation of assessment practice, if there is more than one internal quality assurer for a subject, they should also meet to ensure the consistency of their own practice, systems and record keeping. They can also meet separately from their assessors for their own standardisation activities. A meeting and standardisation plan, as in Table 2.2, will help everyone see when the various activities will take place.

If there is only one assessor and you are the only IQA, the assessor won't have the opportunity to standardise with other assessors. However, in addition to the IQA sampling you have planned to carry out, you could re-assess one or more of the other units. This would enable a discussion to take place to ensure you are both interpreting the requirements in the same way. However, if you have re-assessed a unit, you cannot then internally quality assure it too. Alternatively, your assessor could link up with assessors from other organisations to help standardise their practice, perhaps as part of a consortium which offers the same qualifications.

Sample plan and tracking sheet

A sample plan and tracking sheet will identify what you will sample, from whom and when. You will need to sample a cross-section of work from assessors, learners, methods and decisions. This will be covered in more detail as you progress through the chapter. The plan should track the dates when interim and summative sampling have actually taken place. Interim sampling is ongoing throughout the learner's programme, and summative is at the end. Tracking allows an audit trail to be followed by any external quality assurers or inspectors.

To decide on what to sample and from whom, you should consider the following:

- assessors – qualifications, experience, workload, caseload, locations
- learners – any particular requirements, ethnic origin, age, gender, locations
- methods of assessment and learners' evidence – e.g. observation, questions, witness testimonies, tests, simulation, prior learning, work products
- records and decisions – assessment planning and feedback records, validity and reliability.

Table 2.2 Example meeting and standardisation plan

Meeting and standardisation plan

Qualification: Level 2 Customer Service   IQA: H. Rahl

Activity	Jan	Feb	Mar	Apr	May	Jun	Jul	Aug	Sep	Oct	Nov	Dec
Assessor team meeting	4th		9th		7th		6th		8th		10th	
Assessor standardisation activity		16th Unit 101		13th Unit 102		15th Unit 103		17th Unit 104		12th Unit 105		
IQA team meeting			18th			29th			17th			3rd
IQA standardisation activity				22nd				30th				15th

Once you are familiar with the above, you can plan your sample using an appropriate method. There are various sampling methods you could use; some are more effective than others. The following terminology relates to how they are viewed on a sample plan, for example:

- diagonal (a different aspect from all learners)

- horizontal (different aspects from the same learner over time)

- vertical (the same aspect from different learners)

- percentage (e.g. 10 per cent from each assessor or learner)

- random (unsystematic method but useful as an additional sample if an issue is found)

- theme-based (relating to a particular area of assessment such as work products, questions or witness testimonies).

Table 2.3 shows a mixture of diagonal, horizontal and vertical sampling for five units of a qualification from one assessor and their learners. Two of the units have already been sampled. The internal quality assurer will sample all the evidence provided for each unit. The summative sample ensures all documentation has been correctly completed when the learner finishes their programme. Having the start date and registration date visible will help check that the learners have been registered in a timely manner with an awarding organisation, if applicable. Equally, there should not be a long gap between completion and claiming the certificate. Noting this information helps keep track of how long a learner takes to complete the full programme, compared to other learners. You could use a separate plan for each assessor or combine them for all assessors. It can be a complex process to get a sample plan right; however, it is just a plan and can be amended at any time as necessary.

If any action is required by an assessor as a result of sampling, it's useful to note it on the plan, perhaps by adding the letters AR for action required. In this way, you can quickly see that something needs to be followed up. You could also add a tick or a cross if you noticed any problems with the unit, for example, if the assessor had misinterpreted something. You can then look at your tracking sheet and see at a glance if certain trends are occurring with a particular assessor or if any units need to be standardised.

It is the internal quality assurer's choice what they sample and when, according to their strategy. Don't be influenced to sample what your assessor gives you, as they may be hiding something they don't want you to see.

Activity

State the advantages and limitations of each of the following sampling methods, in respect of the subject you will internally quality assure: diagonal, horizontal, percentage, random, theme-based, vertical. Are there any other methods you could use? If so, what would you hope to gain?

Table 2.3 Example sample plan and tracking sheet

Sample plan and tracking sheet

Qualification: Level 2 Customer Service

Assessor: P. Jones

IQA: H. Rahl

Learner location	Start date Registration date Number	Unit 101	Unit 102	Unit 103	Unit 104	Unit 105	Summative	End date Certification date
Ann Bex X Company	01 Nov 10 Nov 1234ITG	Jan Sampled on 12 Jan AR	Feb Sampled on 18 Feb X AR	Aug	Sept	Oct	Nov	
Eve Holler X Company	01 Nov 10 Nov 1235ITG	Jan Sampled on 12 Jan √	Feb Sampled on 18 Feb X AR					
Terri Frame Y Company	04 Jan 14 Jan 7854URE	Mar		July				
Jon Vanquis Z Company	04 Jan 14 Jan 7855URE	April			Aug			
Naomi Black Z Company	04 Jan 14 Jan 7856URE	July				Sept	Oct	

Key: AR – action required   X – problems noticed in this unit   √ – no problems noticed in this unit

Examples of IQA activities

Once you have your sampling plans, you can begin to carry out various IQA activities. Depending upon your role, these might include:

- sampling assessed learners' work and supporting assessment records

- observing trainer and assessor practice

- talking to learners and/or others

- arranging team meetings

- arranging standardisation activities

- preparing for an external quality assurance visit.

These will now be explained in more detail.

Sampling assessed learners' work and supporting assessment records

An excellent way of monitoring assessor practice and decisions is to sample the learners' work which has been assessed, along with the supporting assessment records. These would include assessment plans, observation reports and feedback records. This should be on an interim and summative basis; interim is part way through each learner's progress and summative is at the completion stage. If a problem is identified at the interim stage, there is a chance to put it right. Summative sampling can check the full assessment process has been completed successfully and that all documents are accurate.

The benefits of interim sampling are that it gives opportunities to monitor (in alphabetical order):

- all assessment types and methods, whether they are robust, safe, valid, fair and reliable

- consistency of decisions between assessors

- consistency of assessor interpretation of what is being assessed

- good practice which can be shared between assessors

- how assessors are completing their records

- how effective assessment planning is

- how effective feedback to learners is

- how learners are progressing and what has been achieved so far

- if any trends are occurring, i.e. learners making the same mistakes or assessors misinterpreting something

- if assessors need any support or further training

- if learners need any support or have any particular requirements

- if there are any problems that need addressing before the learner completes

- the currency, sufficiency and authenticity of learner evidence

- the views of others, e.g. learners, employers and witnesses

- whether the learner has been registered with the awarding organisation (if applicable, as any assessments prior to registration might be classed as invalid by an awarding organisation).

The benefits of summative sampling give opportunities to check (in alphabetical order):

- all documents are fully completed

- all requirements have been met by the learner and the assessor has confirmed this, enabling certificates to be claimed (if applicable) or records of achievement to be issued

- assessors have implemented any action points

- targets and success rates

- the assessment decisions are correct

- the learner evidence is valid, authentic, current, sufficient and reliable.

You should plan to look at all areas of the qualification or criteria being assessed on a continuous basis over time. You should not sample everything from every assessor. Imagine if you were a baker sampling the cup cakes which have been made by several of your cooks. You might only taste one cup cake from one batch made each day by each cook, as tasting them all would render them unsaleable. However, you do need to build up confidence in how your assessors operate, therefore you might begin by sampling more.

Some areas of learners' work that you sample might be incomplete and other areas will be complete. This will enable you to see how the learners are progressing and monitor how long they are taking to achieve the required outcomes. There is no need to sample something from every learner unless there is a reason to do so, for example, a concern about a particular assessor's decision making or a requirement of the awarding organisation.

The dates for your sampling should tie in with the learners' progress and what has been assessed. If your assessors are keeping up to date tracking records of their learners' progress, you should ask to see these. If they are kept electronically, you might be able to access them via a centralised system or you could ask your assessor to e-mail them to you or give you a hard copy.

There's no point planning to sample a particular unit in a certain month if none of the learners is working towards it. Regular communication with your assessors will help you keep track of what is taking place, and your plan can be updated or amended at any time. Team meetings are also a good opportunity for assessors to give you an update of their learners' progress. It is your choice what to sample, not an assessor's; don't ever feel pressured by them to sample something they want you to. There could be a reason why they don't want you to sample something else. Just remember that you are in control. If you find they are having a problem with a particular area, it could be looked at by all the assessors as a standardisation activity.

Whichever methods you choose, they should be fit for purpose and ensure that something from each assessor is sampled over time. You are looking for quality, not quantity. A mixture of diagonal, horizontal and vertical sampling is best as it ensures everything will be monitored over time. Percentage and random sampling are not good practice as aspects can easily be missed. However, they can be used in addition to other methods, such as theme-based sampling, if a problem is found and you need to sample further. If an assessor covers several locations, evidence from learners at each location must be sampled to ensure a standardised approach and fair practice.

Sampling everything from everyone (e.g. 100 per cent sampling) is time consuming and is not good practice. It becomes double assessment rather than quality assurance and doesn't improve or enhance the assessment process.

Once the sample has taken place, a report must be completed and the findings shared with the assessor. An example report can be found in Table 2.9 later in this chapter. The actual date of the sample can then be added to the sample plan and tracking sheet to show an audit trail from planning to completion.

Observing trainer and assessor practice

A good way of ensuring your trainers and assessors are performing adequately is to see them in action. Not only will this give you the opportunity to see what they are doing, but you will also be able to talk to their learners afterwards. Documenting your observations using a checklist like the ones in Tables 2.4 and 2.5 will help to ensure you remain objective when making decisions regarding performance. Your organisation might be inspected by Ofsted if funding is obtained for various programmes. Ofsted inspectors might also observe trainer practice. If this is the case, it would be useful to talk to your staff about their experiences to see if a pattern is emerging with what you have also observed. For example, it could be that some training and development might be required by the staff member as a result.

When arranging to carry out an observation, you will need to make sure the learner is aware that you are not observing them, but their trainer or assessor. Your staff member, and indeed their learner, might be nervous about being observed. You will need to help put them at ease and explain that you are there to help and support, not to be critical of them. You need to ensure they are performing their job role correctly and making valid and reliable assessment decisions. You also need to check that the assessor has made sure the area is safe and that any resources used are accessible and appropriate.

Talking to learners and/or others

After observing your assessor, it's an ideal time to talk to their learner and gain feedback regarding the assessment process. A checklist can be used for this purpose, as in Table 2.6. Always give the learner the opportunity to ask you any questions and to discuss any aspects of the assessment and IQA process with you. If you are not able to answer any questions from the learner, make sure you find out the answer and then get back to them. Not doing so is rude and does not set a good example. Thank the learner for their time and wish them well with their future progress. Don't be tempted to tell them anything about their assessor's performance or make excuses for any problems. You can then give feedback to your assessor away from their learner. This should take place as soon as possible after the observation and in an appropriate location. If you or the assessor has any other commitments at the time, a quick verbal account can be given and then a date and time arranged for formal feedback.

Some of your assessors might also be trainers and you could talk to one or more of their learners after observing a training session. For example, a learner who is an apprentice plumber tells you he is learning about welding copper pipes off the job, but using plastic pipes on the job. It's important for your staff to link off the job training with the on the job training. The learner might feel this is confusing and you will need to talk to the trainer to check that they are communicating with the learner's workplace supervisor. Also, on the job assessment should take place when the learner is ready, not when the assessor is available to do it.

It could be that you identify some areas for development, in which case you will need to discuss this sensitively with your trainer or assessor and reach an agreement on how to proceed. You should always follow up any action points you set to ensure they have been met, and then update your observation checklist accordingly.

When possible, it's useful to take the opportunity to talk to anyone else involved in the progress of learners. This could include witnesses or supervisors from the learner's workplace. This might give you valuable information about the training and assessment process and how the learner is being supported. You should keep a record of any discussions and you might like to adapt one of the checklists in this chapter for this purpose.

Checklists

Your organisation might supply you with checklists to use for some of your sampling activities. This way, you have a formal record of what was carried out and when. If they don't, you could design your own based on those here to ensure you are keeping full and accurate auditable records.

Examples include:

- observation checklist for training delivery (only required if this is part of the process to be monitored for a particular qualification)
- observation checklist for assessment practice
- learner discussion checklist.

The following tables give examples of checklists which you could use. They have been completed as though an internal quality assurer has used them. Records can often be completed electronically and might not need a signature. You should check with your organisation what checklists they provide, if signatures are required, and how they expect you to complete, distribute, save and/or file them. You will find they are similar to ones that EQAs could use. A year should always be added to dates.

Table 2.4 Example observation checklist for training delivery

Observation checklist – training delivery

Trainer: M. Singh

Units/aspects being observed: Unit 101

Number in group: 15

IQA: H. Rahl

Date: 18 February

Checklist	Yes/No N/A	Comments/responses/action required
Were the aims/objectives/learning outcomes clearly introduced?	Y	Learning outcomes were stated and on display which related to the unit
Do learners have action plans/individual learning plans?	Y	All learners have individual action plans to denote the units and target dates for assessment
Is there a scheme of work and session plan?	Y	Both documents were available and relevant to the session being taught – good links from previous to new knowledge
Are the resources and environment safe and suitable?	Y	Training room is suitable, resources are adequate and plentiful for the group size
Does the session flow logically? Are the learners actively engaged?	Y	Topics flowed logically throughout. Learners were involved in group activities, paired and individual work, all were engaged and responsive
Are open questions used to check knowledge?	N	Closed questions were used too often – try and use open questions more
Are all learners able to ask questions and contribute to the session?	Y	Learners were encouraged to ask questions throughout but some dominated over others – try to encourage everyone to ask at least one question
Was learning taking place?	Y	Several different activities were used to observe skills. The knowledge and understanding of each learner was checked by the use of individual questions therefore learning was taking place
Was clear feedback given regarding progress to each learner?	Y	Group feedback was given after each activity, individual feedback was given throughout the session as the opportunity arose
Was the session formally summarised?	N	No summary took place – make sure this happens next time
Was the next session explained? (If applicable)	Y	Next week's topic was clearly explained
Did individual learners know what they had achieved during the session and how it relates to their qualification/programme of learning?	Y	Feedback to each learner linked achievements to the learning outcomes and discussions took place as to how the off the job learning related to the on the job learning
Are all relevant records up to date?	Y	All relevant records were seen for each individual learner, as well as the group profile
Do any learners require additional support?	Y	No learners had disclosed any particular needs – trainer could tactfully ask
Did the trainer perform fairly and satisfactorily?	Y	The session was well planned and delivered and apart from a few action points regarding questioning and summarising, was performed fairly and satisfactorily to ensure learning took place
Does the trainer have any developmental needs?	N	None at the moment

Feedback to trainer:
This was a well planned and delivered session. However, don't forget to involve your learners by asking open questions (e.g. ones that begin with: who, what, when, where, why and how). You need to summarise your session at the end and link to the original learning outcomes.

Table 2.5 Example observation checklist for assessment practice

Observation checklist for assessment practice				
Assessor: P. Jones			IQA: H. Rahl	
Units/aspects being assessed: Unit 101			Location of assessment: X Company	
Date: 31 January				
Checklist	**Yes No N/A**	**Comments/responses/action required**	**Target date**	**Achieved**
Was the learner put at ease and aware of what would be assessed?	Y	You prepared the learner well by explaining what you would observe and how, which helped them relax		
Was an appropriate assessment plan in place?	Y	You had a detailed assessment plan in place		
Were the resources and environment healthy, safe and suitable for the activities being assessed?	Y	All resources were appropriate and suitable for the environment and learner		
Were questions appropriate and asked in an encouraging manner?	Y	You asked open questions to confirm knowledge		
Were current and previous skills and knowledge used to make a decision?	Y	You took into account the fact your learner had already achieved an aspect of the unit previously		
Was constructive and developmental feedback given and documented?	Y	Feedback was very positive and constructive, records were updated		
Was the assessor's decision correct?	Y	Your judgement was correct and accurate		
Did the learner's evidence meet VACSR requirements? (valid, authentic, current, sufficient and reliable)	Y	You checked VACSR for all aspects assessed		
Were all assessment records completed correctly?	N	Complete the observation report – copy to go to your learner	7th Feb	8th Feb
Did the assessor perform fairly and satisfactorily?	Y	Yes, apart from completing the observation report due to time limits		
Does the assessor have any training needs? If so, how can these be addressed?	Y	You asked about taking a refresher First Aid course. Please ask the Human Resources dept for details.	30th April	
Does the assessor have any questions?	Y	Q - Do I need to retake the assessor unit as it's changed? A - No, just keep documents which prove you are meeting the current standards		
What assessment activities were used and why?	Observation and questioning to confirm skills and knowledge			

Feedback to assessor:
I liked the way you were unobtrusive during the observation, yet asked open questions to confirm knowledge as the learner carried out the activity. I appreciate you did not want to fully complete the observation checklist until you had asked a few more questions to confirm knowledge – this was due to time constraints. Once this is done, please ensure you give a copy to your learner.

Table 2.6 Example learner discussion checklist

Learner discussion checklist				
Assessor: P. Jones		IQA: H. Rahl		
Units/aspects assessed: Unit 101		Location of assessment: XY Company		
Learner: A. Bex		Date: 31 January		
Checklist	**Yes No N/A**	**Comments/ responses/ action required**	**Target date**	**Achieved**
Are you aware of your progress and achievements to date?	Y	The learner was able to confirm the units achieved		
Did you discuss and agree an assessment plan in advance?	Y	The learner had a copy for unit 101 and said they always talked to their assessor about the plans		
Do you have a copy of what you are being assessed towards?	Y	Has a full copy of the assessment requirements		
Did you have an initial assessment and/or was your previous skills and knowledge taken into account?	Y	Had told assessor what had been achieved previously		
Were you asked questions to check your knowledge and understanding?	Y	Assessor asks questions when appropriate		
Did you receive helpful feedback?	Y	This was verbal but needs to be formally documented – Assessor to follow up	7th Feb	8th Feb
Is your progress regularly reviewed?	Y	Once every 6-8 weeks		
If you disagreed with your assessor, would you know what to do (i.e. use the appeals procedure)?	N	Assessor to explain the appeals procedure to the learner	7th Feb	8th Feb
Do you have any learning needs or require further support?	N	Learner is happy with support given		
Do you have any questions?	Y	Q – when do I get my certificate? A – when we claim it, which is usually three weeks after successful achievement		

Feedback to assessor:
Your learner was very pleased with the way the assessment was conducted. However, she is unsure of the appeals procedure. Please give her the handout which explains this.

Arranging team meetings

You should have a plan to reflect when you will hold meetings with your team of assessors. If you can plan the dates a year in advance this will ensure everyone knows when they will take place and will therefore be able to attend. If you have a large team you could hold a meeting more regularly, or a smaller team less often. You could use an agenda, like the one in Table 2.7, to ensure all important aspects of the assessment and IQA processes are covered. The items for discussion will differ depending upon the type of programme, whether it is externally accredited and whether funding is received or not.

The agenda could be circulated in advance by e-mail or be uploaded to your organisation's intranet. It might be your responsibility to chair the meeting and take minutes. If so, try and produce them as soon as possible after the meeting. Always ensure everyone who attends, or was absent, receives a copy or can access them electronically. It's important to keep up to date with any changes regarding what is being assessed. If it's an accredited qualification, the content might be revised every few years. Awarding organisations issue regular updates, either by hard copy or electronically. Once you receive these, you need to discuss the content with your team to ensure you all interpret them in the same way, perhaps during the team meeting so that a record is maintained.

Table 2.7 Example agenda

AGENDA **Internal quality assurer and assessor team meeting** **(Date)**
1. Present
2. Apologies for absence
3. Minutes of last meeting
4. Matters arising
5. Programme: recruitment, new starters, leavers, changes to standards and qualifications, feedback from inspections, maths, English, ICT and employability skills
6. Assessment: types and methods used, current progress of learners, issues or concerns, record keeping, continuing professional development (CPD) activities
7. Internal quality assurance – observations and sampling dates, registrations and certifications, appeals and complaints, general feedback to assessors from IQA monitoring activities
8. External quality assurance – feedback and reports, action points, updates from awarding organisations
9. Standardisation – feedback from recent activities, planning new activities and dates
10. Equality and diversity
11. Health and safety, Safeguarding, Prevent Duty
12. Any other business
13. Date and time of next meeting

Arranging standardisation activities

You will need to plan and manage standardisation activities with your assessors and any other internal quality assurers. Standardisation of practice ensures the assessment and IQA requirements are interpreted accurately, and that everyone who is interpreting thinks in the same way and is making comparable, fair and consistent decisions. You could have sepa- rate events for assessors and for internal quality assurers (if there is more than one for your subject). Standardisation events are not team meetings; the latter are to discuss issues relating to the management of the programme, for example, awarding organisation updates, targets, success rates and learner issues. However, if the team is small, you could include a standardisation activity during the team meeting and keep a record of the outcomes.

Example standardisation activities include (in alphabetical order):

- collaborating on creating schemes of work, session plans, course materials and resources for taught courses

- comparing how documents have been completed

- creating a bank of assessment materials, i.e. assignments, multiple choice questions, oral questions, along with expected answers

- discussing any appeals and complaints and how they can be avoided

- discussing how learner evidence can meet the requirements, e.g. the way witness testimonies or learner statements are used

- internal quality assurers agreeing how their practice will be consistent to support their assessors

- interpreting policies and procedures

- interpreting the qualification requirements (or what is to be taught/assessed) and how assessment decisions are reached

- new staff shadowing experienced staff

- peer observations and feedback to ensure consistency of practice

- reviewing assessment activities – looking at safety and fairness, validity and reliability, deciding alternative methods of assessment for any particular learner requirements or needs

- reviewing how assessment plans and feedback records are completed by different assessors

- role play activities such as assessment planning; making a decision; giving feedback; dealing with a complaint

- the way feedback is given to learners

- the way learner reviews are carried out

- updating assessment and IQA documentation, i.e. checklists, records and templates.

One way to standardise practice between assessors is to ask them each to bring along a learner's evidence for a unit or aspect they have assessed, with their supporting assessment plans and feedback records. See Table 2.8 for an example of a completed

Table 2.8 Example standardisation record for assessed work

	Standardisation record for assessed work		
Learner: Ann Bex	Original assessor: J. Smith		
Qualification/unit: Level 2 Customer Service	Standardising assessor: M. Singh		
Aspect/s standardised: Unit 101 evidence and assessment records	Date: 16 February		
Checklist	**Yes**	**No**	**Comments/action required**
Is there an agreed assessment plan with SMART targets?	Y		Your plan had very clear SMART targets with realistic dates for achievement.
Are the assessment methods appropriate and sufficient? Which methods were used?	Y		Appropriate and sufficient methods were used. However, you could reduce the number of observations in the workplace if you can rely on the witness testimonies. Methods used: observation, questioning, products and witness testimonies.
Does the evidence meet ALL the required criteria?	Y		All assessment criteria have been met through the various assessment methods.
Does the evidence meet VACSR?	Y		You have ensured all these points. You also took into consideration an aspect that you hadn't planned to assess, but that naturally occurred during an observation.
Is there a feedback record clearly showing what has been achieved? (Is it adequate and developmental?)	Y		Your feedback is very thorough and confirms your learner's achievements. However, you could be more developmental to guide your learner towards ways of improving her current practice.
Has subsequent action been identified? (if applicable)		N	The feedback record showed what had been achieved and what feedback had been given. However, no further action had been identified. You need to plan which units will be assessed next and set target dates for their achievement.
Do you agree with the assessment decision?	Y		I agree with the decision you made. However, I do feel you could have reduced the number of workplace observations.
Are all relevant documents signed and dated (including countersignatures if applicable)?		N	As you are still working towards your assessor award, you need to ensure your decisions have been countersigned by a qualified assessor.
Are original assessment records stored separately from the learner's work?		N	You have given your original copies to your learner, you need to ensure you keep the original and give your learner the copy. The original must be kept secure for three years in the assessor office.

General comments:

Although there are a few 'No's' in the checklist, this does not affect my judgement as I agree with your decision for this learner. You agreed a SMART assessment plan with your learner which was followed through with assessment and feedback. All your records are in place; however, don't forget to keep originals in the office and give your learner a copy. This is part of our organisation policy due to some learners having amended the original copy in their favour. It's harder to amend a copy as the pen colour is more prominent.

Make sure you set clear targets for future development and assessment opportunities when you give feedback. It's better to do it at this point whilst you are with your learner to enable a two-way conversation to take place and to agree suitable target dates.

As you are still working towards your assessor award, you need to ensure your decisions have been countersigned by a qualified assessor. Please do this by the end of the month.

Comments from original assessor in response to the above:

I agree with your feedback, I had forgotten about keeping original copies and only giving a photocopy to my learner. I will ensure I do this in future. I wasn't able to get hold of my countersignatory as he was on holiday, I will make sure he reads my records and signs them upon his return. I will then take a copy ready to use as evidence for my assessor award. I realise that I must give more developmental feedback and agree future targets when I am with my learner.

Key: SMART: specific, measureable, achievable, relevant, time-bound
         VACSR: valid, authentic, current, sufficient, reliable

standardisation record for assessed work. The learner's work and assessment records can be swapped between assessors who can then re-assess them. The activity could be anonymous if learner and assessor names are removed from the documents beforehand. A discussion can then take place to see if all assessors are interpreting the requirements in the same way and making the same decisions. This is also a chance to see how different assessors complete the records and the amount of detail they write. The activity can lead to an action plan for further training and development of assessor practice, if necessary.

Assessment types and methods should also be standardised at some point. If one assessor has produced a project or an assignment for their learners to carry out, they should share it with the other assessors. This will ensure all learners have access to the same assessment materials. It's also a chance to make sure any questions are pitched at the right level for the learners.

Example

Shane had produced an assignment for his learners which would holistically assess two units from a Level 1 Numeracy qualification. Part of the assignment expected learners to reflect upon their progress. When the assessor team at a standardisation meeting discussed the assignment it was felt that the learners might not be skilled enough to reflect adequately. It was also noticed that an aspect of one of the units had been missed. The team decided that reflection would not be used formally. Shane rewrote the assignment and asked for further feedback from his colleagues before use.

All assessment activities used must be fit for purpose. Formative assessments are often informal and not usually counted towards achievement. They are ideal for checking progress and for planning further training and development. Activities can therefore be designed by the assessor and be more flexible to cater for different learner needs and their progress at the time, for example, using a quiz to check knowledge gained at a given point. Summative assessments are formal and are sometimes produced by the assessor, for example, assignments. Accredited qualifications might have assessments which are produced by the awarding organisation or by the assessor, for example, a test. If more than one assessor is involved for the same subject, they should get together to standardise all activities, resources and expected answers. As an IQA, you will need to check the consistency of assessor practice as part of your monitoring activities.

Learners and employees taking apprenticeship programmes might attend an assessment centre for a summative test. This might be facilitated by someone they have never met before. It is likely to be in a different location to where the training and formative assessment has taken place. You will need to make sure your assessors are preparing their learners for this type of independent assessment. Some workplace assessors might need to learn how to train if this has not been a formal part of their job role.

Benefits of standardisation

The main benefit of standardisation is that it gives a consistent experience for all learners, no matter who their assessor is. It's also a good way of maintaining professional development, and ensuring compliance and accountability with awarding organisations and regulatory authorities' requirements (if applicable).

Other benefits of standardisation include (in alphabetical order):

- an opportunity to discuss changes and developments

- assessment decisions are fair for all learners

- compliance with relevant codes of practice and regulations

- confirmation of own practice

- consistency and fairness of judgements and decisions

- empowerment of teachers, trainers and assessors

- giving staff the time to formally meet

- maintaining an audit trail of aspects standardised

- meeting quality assurance requirements

- re-assessment to find errors or incorrect decisions by assessors, or even plagiarism or cheating by learners

- roles and responsibilities being clearly defined

- setting action plans for the development of assessment activities

- sharing of good practice

- spotting trends or inconsistencies

- succession planning if staff are likely to leave

- upholding the credibility of the delivery and assessment process and practice.

Records will need to be kept of all standardisation activities to show that all units or aspects have been standardised over a period of time.

Using technology for standardisation purposes

Technology can be used for standardisation activities and is ideal if not all the team members can attend a meeting or activity at the same time, or are located in different buildings. When standardising the decisions assessors have made based on electronic evidence, it's important to be sure the work does belong to the learner and that the assessor has confirmed the authenticity of it.

Some examples of using technology for standardisation activities include (in alphabetical order):

- creating, updating and sharing documents online, e.g. schemes of work, session plans, resources and course materials

- holding meetings via Skype or videoconferencing facilities to discuss the interpretation of aspects of a programme or qualification

- making visual recordings of how to complete forms and reports. If a staff member is unsure how to fill in a form they could access a video to see an example

- recording standardisation activities and uploading them to an intranet or virtual learning environment (VLE) for viewing/listening to later

- taking digital recordings or videos of role play activities or case studies, e.g. assessors making decisions and giving developmental feedback. Assessors could view them remotely to comment on strengths and limitations of a particular method

- using online webinars to standardise delivery and assessment approaches.

Activity

Plan an activity to standardise an aspect of practice between your assessors. Carry out the activity and evaluate how well it was received and carried out by your assessors. Decide what could be done differently next time.

Preparing for an external quality assurance visit

If you are internally quality assuring an accredited qualification or endorsed programme from an awarding organisation (AO), you will be visited by an external quality assurer (EQA) at some point. How often these visits take place, the duration and the activities carried out will depend upon the requirements of the qualification and how active your organisation is. The EQA or a representative from the AO will make contact to arrange a suitable date and time for the visit. There will be timescales for the EQA to follow when contacting you for information and sending you their visit and sample plan. They will probably have formal documentation which they will send to you which will outline the information they require prior to the visit. This might include:

- a list of trainers, assessors, witnesses, IQAs and their locations

- copies of minutes of meetings and records of standardisation activities

- details of learners, their locations and registration and/or unit completion and certification dates.

The EQA will use this information to plan what they want to see and from whom. They will be able to compare your learners' names and registration/certification dates with those on the AO's database. If any information is different, this will highlight an anomaly for them to check with you.

Once the EQA has the relevant information, they will send you a *visit and sample plan*. An example EQA visit and sample plan can be seen in Table 4.2 in Chapter 4. This will show what they will want to achieve during their visit and might include:

- looking at assessment, IQA and other supporting documents and records

- meeting the team

- observing assessor and IQA practice

- sampling learners' work

- talking to learners and witnesses.

The EQA will not always want to see completed learner work. They might request to see learners' work which has been formatively as opposed to summatively assessed, to check ongoing progress. They might also ask for assessed work which has been sampled by you as the IQA on an interim rather than a summative basis. They should sample the same unit from different assessors to ensure consistency and sample across all units, assessors and locations of assessment. A variety of assessment methods, evidence and records might also be sampled.

If this is the first visit by the EQA, they may require a lot more information than a routine visit. However, they should convey what they need to see in advance. There will be certain things you will need to prepare, and you should involve your team members to help you fulfil these. You should refer to any previous EQA reports to ensure all action points have been met. If any have not been met, you might be given a sanction. This means certification rights could be removed and therefore you can no longer claim certificates for learners who have successfully completed.

The following checklist, which is in alphabetical order, lists the documents you should have available (if relevant) prior to the EQA visit. Records can be electronic or manual, but they must follow data protection and confidentiality requirements. Whilst your EQA might not wish to look at all of the following, it's useful to be prepared.

Checklist – documents for an EQA visit

☐ *Action plans, assessment plans or individual learning plans (ILPs) for each learner*

☐ *Assessment tracking sheet showing all assessors, all learners and dates of achievement*

☐ *Assessor & IQA induction materials*

☐ *Copies of registration and certification details for all learners*

☐ *Equal Opportunities/Equality and Diversity policy and monitoring data*

☐ *Evidence of completed action points from the previous EQA sample*

☐ *Evidence of learner support/particular assessment requirements being met*

☐ *Feedback from learners, for example, evaluations and survey results, along with actions taken*

☐ *Identification checks and learner authenticity statements confirming the work is their own*

(Continued)

- [ ] *Initial and diagnostic assessment records for each learner*

- [ ] *IQA planning and tracking sheet*

- [ ] *IQA records: interim and summative sample plans and tracking sheets, observations of assessors, discussions with learners and others such as witnesses*

- [ ] *IQA policy, rationale and strategy*

- [ ] *Learner induction materials and learning support details*

- [ ] *Learner portfolios with copies of assessment plans, decisions and feedback given. Separate records should be maintained of original documents. The EQA will want to know how you store your records securely. Extra portfolios should be available in case the EQA decides to carry out an additional random sample*

- [ ] *List of assessment sites/locations and work placements*

- [ ] *List of IQAs/assessor names and contact details*

- [ ] *Minutes of assessor/IQA meetings*

- [ ] *Organisation chart*

- [ ] *Overall learners' start dates, registration dates, unit completion dates, end dates, certification dates*

- [ ] *Previous EQA reports and evidence to show actions completed*

- [ ] *Records of any appeals and complaints*

- [ ] *Records of standardisation activities*

- [ ] *Relevant policies and procedures*

- [ ] *Reviews of progress and/or tutorial records*

- [ ] *Staff CVs, certificates and CPD logs*

- [ ] *Training materials i.e. scheme of work, session plans, resources and a list of equipment used by the learners*

To help you prepare for the visit, you could refer to the report form that the EQA will use on the day. If you've been visited before, you will be able to look back at the last report. If not, you could ask the EQA if you can see an example report form. You might like to meet with your assessors in advance of the visit, look at a sample of their work and use the report form as a trial activity to see if you meet all the points. You should ensure you have evidence of meeting any action points which were stated on the previous report. If you are asked something by an EQA and you answer 'yes', make sure you have the documentation or electronic records to prove it.

Other aspects to prepare in advance include informing reception staff of the name, date and time of arrival of the EQA, and arranging a suitable room for the duration of the visit. You might also need to communicate with the EQA regarding transport and/or parking arrangements. All the required documents should be placed in the room (or be accessible) and all requested staff should be available as needed. You should remain professional throughout the visit and be helpful regarding the EQA's requests. If you think you have been asked to do something which you feel is not relevant, you will need to be confident enough to challenge the EQA. They can only ask you to do what is required by the awarding organisation, the regulators and any relevant guidelines and legislation for your subject.

Example

Roisin, the EQA, asked that a particular unit of the qualification should have three observation reports from the assessor. Harry, the IQA, knew that only one was needed. He therefore asked Roisin to show him where it said that three were needed. When she looked at the qualification specification, she noticed Harry was right in that only one was needed. An EQA can make recommendations for improvement but cannot force you to do something which is not a requirement.

You might find the visit will be quite tiring as you will be asked lots of question. You might also need to escort the EQA to different locations to meet staff and learners. Conversely, the EQA might want to be left alone when carrying out their sample. At the end of the visit, the EQA should talk you through their findings and the content of their report. They should discuss and agree any action and improvement points with you. Action points will be sanctionable if not met, whereas improvement points are just for development. The EQA might leave you with a copy of their report, or this might be sent to you via the AO afterwards, or be accessible online. Reading Chapters 3 and 4 of this book will help you understand the role of the EQA and help you prepare for other activities such as a systems visit or a remote monitoring activity.

Extension activity

Find out when the next EQA visit is due for the qualification you are responsible for. Use the previous checklist to help you prepare for it. Make a list of any questions or points that you would like clarifying by the EQA either before or on the day of the visit.

Making decisions

A big part of your role will be to make decisions as to whether your assessors are performing their role competently. Besides observing them in practice, you can sample their assessment planning and decisions. When sampling their learner's work, you are not re-assessing or re-marking it but making a decision as to whether it meets the assessment requirements. You are ensuring the assessor's plans and feedback records are documenting all the activities and that their decisions are valid and reliable. You might agree with your assessor, in which case you can give them feedback as to what they have done well. You might agree but feel your assessor could have given more developmental feedback to their learner. You can then give your assessor appropriate feedback as to how they could develop and improve their feedback skills. Alternatively, you might disagree with an assessor's decision and refer the work back to them. If this is the case, you would need to be very explicit as to why the work had not met the requirements and give your assessor advice on how they can support their learner's achievement. This should all be documented and a target date for completion agreed. Table 2.9 is an example of a completed internal quality assurance sample report. Make sure you keep a note on your sample plan and tracking sheet (as in Table 2.3) if any action is required, so that you can follow it up by the target date. Always update your records to reflect what was sampled and when, and what action has been met and when. A clear audit trail of all activities must be maintained to assist compliance and transparency. It also helps in the event of an appeal or a complaint.

When sampling learners' work, you should always read the accompanying assessment plans and feedback records and any other assessment records such as observation reports and witness testimonies. Reviewing all the assessment documentation will help you gain a clear picture of learner progress and achievement. If witnesses are used, you should contact a sample of them to confirm their authenticity and that they understand what their role entails. If learner achievement relies heavily on the use of witness testimonies as evidence, then you, or someone else, might need to carry out adequate training and give support to the witnesses.

You should compare different assessors' records to ensure they are completing them in a standardised way. Some assessors might be very brief with their feedback and others quite comprehensive. If this is the case, then the assessor who only writes brief comments might not be fully supporting their learner. You could take copies of some records, remove the assessors' names and use them during a standardisation meeting to agree a consistent approach.

Assessors should never do any work for their learners but help them see how they can achieve it for themselves. If you have a concern when sampling the work of one assessor, you could sample more work from them for the same aspect but with different learners to see if it was a one-off. You could then sample the same unit from different assessors to see if there is a trend. If you find that other assessors are also having the same problems with the same aspect, you should contact them all straight away. It might be that they have all misinterpreted the same thing; therefore you need to update them immediately. Otherwise, the learners might be disadvantaged due to no fault of their own. You could then schedule this particular area for standardisation. This would enable everyone to discuss their interpretation of the criteria, to share ideas and support new/inexperienced staff as necessary.

When sampling work from different assessors, if you are sampling the same aspects you can see how consistently the different assessors are performing. You can then note any inconsistencies to discuss at the next team meeting. For example, if one assessor is giving more support to learners or expecting them to produce far more work than others, then this is clearly unfair.

Table 2.9 Example internal quality assurance sample report

Internal quality assurance sample report	
Learner: A. Bex	Programme/Qualification: Customer Service Level 2
Assessor: P. Jones	IQA: H. Rahl
(Interim)/Summative sample	Date: 12 January

Unit/aspect sampled	Comments
Unit 101	I have observed the assessor performing satisfactorily today – see separate observation checklist for full details. An assessment plan had been agreed between the assessor and learner but had not been dated. Products of work, a witness testimony and a learner statement were sampled. The feedback record completed by the assessor regarding the evidence which had already been assessed is detailed and constructive. The assessor just needs to fully complete the observation checklist, and document the questions and answers asked today.

Is the evidence?	If no – action required and target dates:	Date action completed:
Valid (Yes)/No		
Authentic (Yes)/No		
Current (Yes)/No		
Sufficient (Yes)/No		
Reliable (Yes)/No		
Have assessment plans and records been completed, signed and dated?		Yes/(No)
Is the assessor's decision correct?		(Yes)/No
If summative IQA – can the certificate now be claimed?		Yes/No/(NA)

Feedback to assessor:	Assessor's response:
Don't forget to date the assessment plan, and complete and sign your observation checklist. Give a copy to your learner by 7th February.	I plan to complete this by next Friday.

Sampling learners' work is also a good opportunity to check for aspects such as plagiarism and copying. It could be that two learners have submitted a piece of work which is almost identical. Hopefully your assessor will have noticed this. However, you would need to satisfy yourself that they hadn't worked together or copied one another's work. You and your assessors need to be aware of learners colluding or plagiarising work, particularly now that so much information is available via the internet. Learners should take responsibility for referencing any sources, if applicable, and may be required to sign an authenticity statement to confirm the work is theirs. If you suspect plagiarism, you could type a few of their words into an internet search engine or specialist program and see what appears. You would then have to refer it back to your assessor to challenge their learner as to whether it was accidental or intentional.

Activity

Find out what your organisation's policy is regarding cheating, copying and plagiarism, and find out what your involvement would be. Ensure all your assessors are aware of the policy and discuss any issues at your next team meeting.

When sampling, you need to make sure all work is valid, authentic, current, sufficient and reliable (see Chapter 1 for details). If you are in any doubt, you must refer it back to your assessor or have an informal discussion with them. It might be that the assessor has omitted to state something on their feedback record or it might be more serious. Never feel pressured to agree with your assessor if you feel something is not quite right. If there are other internal quality assurers in the same subject area as yourself, you could discuss your findings with them first.

You should also be making decisions as to whether the assessment types and methods used are adequate, fair and appropriate. If one assessor is carrying out three observations with all their learners, but another is only carrying out one, then that is clearly not fair. If one assessor is expecting some learners to write a 2,000 word essay, and others a 1,500 word essay, then again, that's not fair. Your assessor might justify their actions by saying they are challenging the more able learners; however, unless the assessment requirements state this is acceptable, then it isn't. It's fine to differentiate formative assessments but summative assessments need to be fair to all learners.

As you are only sampling aspects of the assessment process, there will be some areas that get missed. This is a risk as you can't sample everything from everyone. You need to build up your confidence in your assessors to know they are performing adequately. If you find a problem when sampling, or have any concerns, you will need to increase your sample size. Conversely, you could reduce the sample size for your experienced assessors if you have confidence in them. However, never assume everything is fine as experienced assessors could become complacent. If all your assessors know what you will be sampling and when, they might not be as thorough with the areas you are not due to sample. You can always carry out an additional random sample at any time and ask to see an aspect of assessor practice which isn't part of your original plan. Your plan is a live document which can be amended at any time.

After a period of sampling, you should analyse your findings and give overall feedback to your assessors, perhaps at the next team meeting. You might have found patterns or trends, for example, all assessors are making the same mistake within a particular unit. You might see that one assessor is taking more time than others to pass their learners, or that several of their learners have left. Conversely, you might find another assessor whose learners are completing really quickly and you will need to find out why.

Appeals and complaints

An appeal is usually about an assessment decision, whereas a complaint is more likely to be about a situation or a person. Learners who appeal or complain should be able to do so without fear of recrimination. Confidentiality should be maintained where possible to ensure an impartial outcome, and the learner should feel protected throughout the full process. If anyone does make an appeal or complaint, this should not affect the way they are treated and the outcome should not jeopardise their current or future achievements.

At some point during the assessment process, a learner may wish to appeal against one of your assessor's decisions. There should be an appeals procedure with which learners and assessors are familiar. Information could also be displayed on noticeboards, in a learner handbook, or be available via your organisation's intranet or virtual learning environment (VLE). Learners will need to know who they can go to and that their issue will be followed up. This process will involve various stages and have deadlines, such as seven days to lodge an appeal, seven days for a response etc., and all stages should be documented. Usually, an appeals process is made up of four stages: assessor, internal quality assurer, manager and external quality assurer (if applicable). It might be your role to monitor appeals from learners and it could be that at the second stage in the process you will need to make a decision whether to uphold the learner's appeal. If you do not uphold it, it will then escalate to the next stage, i.e. your manager. The manager's decision could be final, or if the qualification is externally quality assured, it might then escalate to the external quality assurer, whose decision should be final. The nature of the appeal can be used to inform future practice and to prevent further appeals.

Example

Cheng had lodged a formal appeal to the internal quality assurer, Alison, regarding his assessor's decision for unit 301. He felt he should have passed as he had supplied all the required evidence. Alison spoke to the assessor and reviewed Cheng's evidence. It transpired that the assessor was correct in asking for a further piece of evidence. Alison spoke to Cheng and explained that four pieces of evidence were required and he had only submitted three. He accepted Alison's decision and agreed to supply a further piece of evidence. Cheng then retracted his appeal.

If Cheng had used the first stage of the process and discussed his concern with the assessor, there would have been no need to involve the internal quality assurer. Some organisations will provide an appeals' pro-forma for learners to complete, which ensures all the required details are obtained, or encourage an informal discussion with the assessor first. Statistics should be maintained regarding all appeals and complaints; these will help your organisation when reviewing its policies and procedures, and should be provided to relevant external inspectors if requested.

A complaint could also be made by one of your assessors against something you have or have not done as part of your IQA role. You could have an assessor appeal against a decision you have made, for example, if you disagreed with one of their judgements. There should be a formal appeals procedure for assessors, similar to the one for learners. However, if possible, try and discuss it with your assessor first to reach an amicable outcome.

Having a climate of respect and honesty can lead to issues being dealt with informally, rather than procedures having to be followed which can be upsetting for both parties concerned.

Activity

Locate, read and summarise your organisation's policies and procedures for appeals and complaints in your particular subject area. Are there any templates or documents supplied and what are the time limits for submission and follow up? If your qualification is accredited, find out what role the awarding organisation plays regarding appeals and complaints.

Communicating with assessors

Communication is the key to productive working relationships. People act differently depending upon the situation they are in and the people they are with at the time. You might find that on a one-to-one basis an assessor is quite mature but in a meeting they can be rather disruptive and immature. This section will cover some theories of communication, which might help you improve your working relationships.

Transactional Analysis

Berne's (1973) Transactional Analysis Theory is a method of analysing communications between people. Berne identified three personality states: the *child*, the *parent* and the *adult*. These states are called *ego states* and people behave and exist in a mixture of these, due to their past experiences, gestures, vocal tones, expressions, attitudes, vocabulary and the situation they are in at the time.

Transactional Analysis assumes all past events, feelings and experiences are stored within and can be re-experienced in current situations. You might see this with assessors who take on a different state depending who they are with. For example, acting like a child and asking for help from a colleague, but acting like an adult with a manager.

Transactions are verbal exchanges between two people: one speaks and the other responds. If the conversation is complementary then the transactions enable the conversation to continue. If the transactions are *crossed*, i.e. child to adult, rather than adult to adult, the conversation may change its nature or come to an end.

Berne recognised that people need *stroking*. Strokes are acts of recognition which one person gives to another and can be positive or negative, i.e. words of appreciation or otherwise. Giving or receiving positive strokes develops emotionally healthy people who are confident in themselves and have a feeling of being *okay*. Negative strokes can lead to a person being demoralised if not given in a skilful way.

Example

Ibrahim was working towards his assessor qualification and wanted to prove to his assessor how good he was. He kept saying to himself: I'll be okay if I produce all the required work to please my assessor and I don't make any mistakes. By doing this, Ibrahim felt he would be looked upon more favourably and receive strokes of appreciation in the form of positive feedback, which he felt would encourage and motivate him.

Understanding a little about the different states of the child, parent and adult will help you see how your assessors take on different roles in different situations, particularly in meetings where some may be more vocal than others.

If you ever feel like a *child* at work, it may be because your manager is operating in their *parent* mode and you are responding in your *child* mode. Your *child* makes you feel small, afraid, undervalued, demotivated and rebellious. These feelings may make you undermine, withdraw, gossip, procrastinate, plot revenge or attempt to please in order to be rewarded. In this *child* mode, you will find it very hard to become a successful professional.

As you are in a managerial role, you may find yourself acting like a *parent*. You may have learnt this from your parents' responses to you years ago. The *parent* mode makes you feel superior, detached and impatient. Being in this state can make you harden your tone, not listen to people, shout, bribe others into complying and criticise them more than encourage them.

The best option is to remain in the *adult* state. As an *adult*, you feel good about yourself, respectful of the talents and lives of others, delighted with challenges, proud of accomplishments and expectant of success. These feelings make you respond to others by appreciating and listening to them, using respectful language, perceiving the facts, considering alternatives and having a long-term view and enjoyment of work and life.

If you realise that you have moved into a *role*, it is possible to change if you need to. When you feel your *child mode* about to make you withdraw, gossip or undermine, you can choose instead to participate, find out the facts and resolve your differences in the *adult* state.

When you feel your *parent mode* about to make you criticise or take over, you can choose instead to speak warmly, be patient, listen and find enjoyment in the challenge.

However, it is very difficult to consistently be in the *adult* state. You may find yourself adapting to different situations and responding and reacting to the states other people have taken on. However, trying to remain in the adult state should help you gain confidence and respect from your assessors, as well as perform your job role satisfactorily.

Belbin's team roles

Belbin (2010: 23) defined team roles as: *a tendency to behave, contribute and interrelate with others in a particular way.* Belbin's research identified nine clusters of behaviour, each of which is termed a *team-role*. Each team-role has a combination of strengths they contribute to the team and allowable weaknesses. It's important to accept that people have weaknesses; therefore if you can focus on their strengths you will be able to help manage their weaknesses. See Table 2.10 Team Role Summary Descriptions for the contributions and allowable weaknesses of each team role.

Table 2.10 Team role summary descriptions, reprinted with kind permission from www.belbin.com

Team Role	Contribution	Allowable Weaknesses
Plant	Creative, imaginative, free-thinking. Generates ideas and solves difficult problems.	Ignores incidentals. Too preoccupied to communicate effectively.
Resource Investigator	Outgoing, enthusiastic, communicative. Explores opportunities and develops contacts.	Over-optimistic, Loses interest once initial enthusiasm has passed.
Co-ordinator	Mature, confident, identifies talent. Clarifies goals, Delegates effectively.	Can be seen as manipulative. Offloads own share of the work.
Shaper	Challenging, dynamic, thrives on pressure. Has the drive and courage to overcome obstacles.	Prone to provocation. Offends people's feelings.
Monitor Evaluator	Sober, strategic and discerning. Sees all options and judges accurately.	Lacks drive and ability to inspire others. Can be overly critical.
Teamworker	Co-operative, perceptive and diplomatic. Listens and averts friction.	Indecisive in crunch situations. Avoids confrontation.
Implementer	Practical, reliable, efficient. Turns ideas into actions and organises work that needs to be done.	Somewhat inflexible. Slow to respond to new possibilities.
Completer Finisher	Painstaking, conscientious, anxious. Searches out errors. Polishes and perfects.	Inclined to worry unduly. Reluctant to delegate.
Specialist	Single-minded, self-starting, dedicated. Provides knowledge and skills in rare supply.	Contributes only on a narrow front. Dwells on technicalities.

The nine team-roles are grouped into *action*, *people* and *cerebral* roles:

- action-oriented roles: shaper, implementer and completer finisher

- people-oriented roles: co-ordinator, team worker and resource investigator

- cerebral roles: plant, monitor evaluator and specialist.

Sometimes groups or teams become problematic, not because their members don't know what they are doing, but because they have problems accepting, adjusting and communicating with each other as they take on different roles. Knowing that individuals within teams take on these different roles will help you manage group work more effectively. For example, if a group of assessors needs to work together on a project, you could have a mixture from the *action*, *people* and *cerebral* roles within the group.

Emotional intelligence

Emotional intelligence (EI) is a behavioural model, given prominence in Daniel Goleman's book *Emotional Intelligence* (1995); however, work originally began on the model in the 1970s and 1980s by Howard Gardner.

Goleman identified five domains of emotional intelligence:

- knowing your emotions

- managing your emotions

- motivating yourself

- recognising and understanding other people's emotions

- managing relationships, i.e. the emotions of others.

The principles of EI provide a new way to understand and assess people's behaviour, attitudes, interpersonal skills, management styles and potential. This could be useful if you have a large team of assessors who aren't always working in a consistent manner. By developing emotional intelligence Goleman suggested that people can become more productive and successful at what they do, and help others to be more productive and successful too. He also suggested that the process and outcomes of developing emotional intelligence contain aspects which are known to reduce stress for individuals and organisations. This can help improve relationships, decrease conflict, and increase stability, continuity and harmony.

Becoming aware of your own emotions and how they can affect your activities will help you develop more fulfilling and professional relationships with your assessors.

The EI concept argues that IQ (Intelligence Quotient), or conventional intelligence, is too narrow as there are wider areas of emotional intelligence that dictate and enable how successful people are. Possessing a high IQ rating does not mean that success automatically follows.

Neuro Linguistic Programming

Neuro Linguistic Programming (NLP) is a model of interpersonal communication concerned with relationships and experiences. Interpersonal skills are those which take place *between people*, in contrast to intrapersonal skills which are *within a person*.

The model can be useful when providing feedback to assessors to help influence their development. NLP is a way to increase self-awareness and to change patterns of mental and emotional behaviour. Richard Bandler and John Grinder, the co-founders of NLP in the 1970s, claimed it would be instrumental in finding ways to help people have better, fuller and richer lives. They created the title to reflect a connection between neurological processes (neuro), language (linguistic) and behavioural patterns (programming) which have been learned through experience and can be used to achieve specific goals. The model was based on how some very effective communicators were habitually using language to influence other people.

Activity

Arrange to observe an experienced internal quality assurer, preferably in your subject area, to watch how they interact with their assessors. Are they using any NLP techniques to make communication easier? If you can't carry out an observation, watch or listen to influential people on the television, radio or the internet. Ask yourself what it is about them that makes them successful. Can you emulate this in yourself? If so, how?

NLP training should help turn negative thoughts into positive thoughts. It provides the skills to define and achieve outcomes, along with a heightened awareness of the five senses.

NLP techniques can be used to:

- coach assessors how to gain greater satisfaction from their contributions
- enhance the skills of assessors
- improve own and others' performance
- improve an individual's effectiveness, productivity and thereby profitability
- set clear goals and define realistic strategies
- understand and reduce stress and conflict.

NLP provides questions and patterns to make communication more clearly understood, for example, all thoughts and behaviours have a structure, and all structures can be re-programmed. Do you use jargon, complex terms or clichés without thinking? Your assessors might not understand what you are talking about because you assume they already have the knowledge. To improve your own communication skills, you could observe others who are skilled and experienced to see how they perform.

Demonstrating and evaluating your interpersonal and intrapersonal skills should help you deal effectively with situations which might occur with your assessors.

Extension activity

Hold a team meeting or standardisation activity with your team of assessors. Afterwards, evaluate the interactions among everyone. Was anyone acting differently to how they normally would behave? Did you notice anyone taking on any traits of Belbin's roles, Berne's personality states and/or using NLP and EI principles? Would you have done anything differently based on your knowledge of these theories? You might like to research these theories further and others if you are interested in the way people interact with each other.

Providing feedback to assessors

You should provide feedback to your assessors whenever you get the opportunity. Formal feedback could be given verbally after an observation of their practice, after sampling their assessed work or during an appraisal. This should always be followed up with written feedback. Informal feedback can be given at any time to help confirm an assessor's practice. It can enable your assessors to see what they are doing right and what they can do to improve. Formal feedback should be given at an appropriate date, time and place, and in a constructive and developmental manner. Always remember that it is unprofessional to give feedback to an assessor in front of their learners. You should also give your assessor a copy of any report you have completed. This acts as a formal record of feedback and any action required, which should always be followed up.

Activity

Think back to the last time you gave feedback, to either an assessor or someone else. Was it informal or formal? Did you keep a record? How did you feel afterwards and how did the other person react to what you said? Is there anything you would do differently? If so, why?

Feedback should be concentrating on the assessment process and not be critical of the assessor as a person. It should be used to confirm competence, and to motivate and encourage rather than apportion blame for any reason. Above all, it should be to help your assessor develop their assessment practice and to maintain and improve the quality of the assessment process for their learners.

If you find something that the assessor has done wrong, or that they could improve upon, don't be critical but state the facts. You could ask your assessor to reflect upon their performance before you give feedback. That way, they might realise any mistakes before you have to point them out. You can then suggest ways of working together to put things right. It could be that your assessor was unaware of something they should or should not have done. Communicating regularly with them and identifying any training needs could prevent problems from occurring.

You might not find anything wrong, in which case you still need to give feedback, which will confirm that what they are doing is right. If you have an assessor who is performing really well, you could ask them to mentor an underperforming or new assessor, providing they have the time.

Always allow your assessor time to clarify anything you have said and to ask any questions. Don't interrupt them when they are speaking and avoid jumping to any conclusions. Use eye contact and listen carefully to what they are saying. Show that you are a good listener by nodding your head and repeating key points. Your assessor should leave knowing exactly what needs to be done and by when.

Example

Sureya sampled six assignments that Joey had marked from his group of 30 learners. She found his handwritten records difficult to read and his feedback to his learners was very sparse. However, all the learners had met the criteria and produced some really good work. When Sureya gave feedback to him, she began with something positive and then moved on to the developmental points: 'Joey, I enjoyed reading your learners' assignments and I felt they had all put in a lot of effort. However, I do feel that your feedback to them is not as detailed as it could be and I found your writing hard to read. Could you word-process your feedback and take the opportunity to state something specific regarding each learner's achievement next time?'

This example shows how Joey could improve his feedback skills to help his learners. It was also given in an encouraging manner. The use of the word *however* to link the points is much better than the word *but*, which can sound negative.

Feedback should always be (in alphabetical order):

- based on facts, not opinions, and aimed at assessors not learners
- clear, genuine and unambiguous
- developmental – giving examples for improvement or what could be changed
- documented – records must be maintained
- focused on the activity not the person
- helpful and supportive – guiding the assessor to useful resources and CPD activities
- honest and detailed regarding what was or wasn't carried out
- positive and constructive – focusing on what was good and how practice can be improved or changed
- specific and detailed regarding what was sampled and what was found
- strategic – seeking to improve performance.

Different feedback methods

There are many ways you can give feedback formally or informally. These include being:

- descriptive – describes examples of what was done well, what could be improved and why

- evaluative – usually just a statement such as *well done* or *good*. This method is not descriptive and does not offer helpful or constructive advice. It does not give people the opportunity to know what was well done, what was good about it or how they could improve. It's just an evaluation of achievement which doesn't offer detailed feedback

- constructive – is specific and focused to confirm actions or to give developmental points in a positive and helpful way

- destructive – relates to improvements which are needed and is often given in a negative way which could demoralise the person

- objective – clearly relates to specific requirements and is factual regarding what has and has not been met

- subjective – is often just a personal opinion and can be biased, for example, if the IQA is friendly with the assessor. Feedback given subjectively might be vague and unhelpful.

When giving feedback to people, you need to be aware that it could affect their self-esteem. The quality of feedback received can be a key factor towards their development. Ongoing constructive feedback which is developmental and has been carefully thought through is an indication of your interest in the person and of your intention to help them develop and do well in the future.

When giving feedback:

- own your statements by beginning with the word *I* rather than *you* (however, written feedback could be given in the third person if your organisation prefers)

- start with something positive, for example, *I really liked the confident manner in which you gave feedback to your learner*

- be specific about what you have seen, for example, *I felt the way you used those assessment methods were just right for your learner*

- offer constructive and specific follow-on points, for example, *I feel you could have informed your learner where they could locate the particular document you were talking about*

- end with something positive or developmental, for example, *I liked the way you prepared for the observation today, you were well organised and your learner was able to achieve what you expected of them.*

Being constructive, specific and developmental with what you say and owning your statements by beginning with the word *I* should help your assessor focus upon what you are saying and listen to what you say to help them improve. If you don't have any follow-on points then don't create them just for the sake of it. Conversely, if you do have any negative points or criticisms, don't say *My only negative point is …* or *My only criticisms are ….* It's much better to replace these words and say *Some areas for development could be …* instead.

Extension activity

Carry out an activity such as sampling some learners' work which has been assessed. Did you agree with your assessor's decision? How detailed was their feedback to their learner? What feedback would you give to your assessor and why?

Record keeping

It is important to keep records to satisfy your organisation's requirements, as well as those of awarding organisations, regulatory authorities and funding bodies. This will usually be for a set period, for example three years. If you are ever unsure as to when to make a record of something, just say to yourself *if it isn't documented, there's no proof it took place.* You always need to prove what did, as well as what did not happen, by keeping full and accurate records.

Reasons for keeping IQA records include (in alphabetical order):

- in case of a complaint or an appeal against an assessment or IQA decision

- in case of malpractice, e.g. plagiarism by a learner

- to document equality and diversity data

- to meet your own organisation's requirements, as well as those of an awarding organisation and/or regulatory authorities and funding bodies

- to monitor assessors' progress and development

- to prove action points from internal and external inspections have been met

- to prove IQA activities have been carried out correctly

- to show that feedback from questionnaires and surveys has been taken into account when reviewing the programme.

Records should always be accurate and based on what you have seen. They should be legible and kept safe, secure and confidential. Most records can be kept manually or electronically. Backing up your data and records is important, particularly if you have electronic records, for example in the case of power failures.

Example

Bill knew that he was due a visit from his external quality assurer next month and wanted to start preparing for it. As he is employed part time, he keeps his records electronically on his own laptop. However, a virus had recently been detected and he realised his records should be kept on the company's computer system, and not just on his. He checked his records were virus free, copied them over, protected them with a password and then deleted them from his own laptop. He also ensured a backup copy would be automatically created at the company.

There might be a standardised approach to completing the IQA records, for example, the amount of detail which must be written or whether the records should be completed manually or electronically. You will need to find out what your organisation expects you to do. If hard copies are used as opposed to electronic ones, the original records should be kept, not photocopies or carbon copies, to guarantee authenticity.

Activity

Make a list of the records and documents you will need to keep as part of your IQA role. Compare this to the list in Table 2.11 on page 74. Add any others you can think of. Find out where you will obtain the documents, whether they are manual or electronic, and which version they are, along with where and how you should maintain them. If there is no formal record-keeping system currently in place, you may need to design your own.

Records should always be accurate, detailed, dated and legible. If you are sampling hard copies of learners' work and assessment records, some AOs prefer the use of a coloured pen to date and initial what you have looked at. This would create a visual audit trail which is easily identifiable if an external quality assurer wishes to see what you have sampled. The dates you use against your initials should agree with the dates on your reports and tracking sheets.

When completing any records, any required signatures should be obtained as soon as possible after the event if they cannot be signed on the day. Any signatures added later should have the date they were added, rather than the date the form was originally completed. If you are completing documents electronically, you will need to find out if an e-mail address, scanned or electronic signature is required or not. This might be acceptable providing the identity of the person has been confirmed and a record kept of the original signature.

Besides keeping your own records in a secure system, you must ensure your assessors are keeping the required records for their role. All records should be kept confidential, be in a secure system and should only be accessible by relevant staff. You also need to ensure you comply with organisational and statutory guidelines such as those for confidentiality, the Data Protection Act 1998 and the Freedom of Information Act 2000 as mentioned in Chapter 1.

Table 2.11 Examples of internal quality assurance records and documents

IQA records and documents	
• assessor observation checklists • details of assessors and IQAs, e.g. contact details, their CVs, CPD records and copies of certificates • enrolment, retention, achievement and progression data • equality and diversity data • external quality assurance reports • internal quality assurance sample reports (interim and summative) • interviews with learners, witnesses and others • IQA rationale and strategy for your subject area	• inspection reports and actions taken • learner discussion checklists • minutes of meetings • observation plan, meeting and standardisation plan, sample plan and tracking sheet • questionnaire or survey results with evaluations and actions taken • records of appeals and complaints • records of learner registration and certification details with an awarding organisation (if applicable) • self-assessment reports • standardisation activity records

Extension activity

What internal organisational or external regulatory requirements must you follow regarding record keeping and why? What impact will they have on your role and that of your assessors? Research the requirements of the Data Protection Act 1998 and the Freedom of Information Act 2000 if you are not familiar with them.

Evaluating practice

Evaluation is not another term for assessment. Evaluation relates to the programme, whereas assessment relates to the learners. Assessment is specific towards a learner's progress and achievement and how they can improve and develop. The evaluation process should aim to obtain feedback from learners, assessors and others to help improve and develop practice, and the overall learner experience on the programme.

Evaluation includes surveys or questionnaires, reviews, appraisals, and informal and formal discussions, telephone calls and meetings. Feedback can come from learners, assessors, internal and external quality assurers, and others such as witnesses and employers. Information gained from evaluations should lead to an improvement for the learners, your assessors, yourself and your organisation. Never assume everything is going well just because you think it is, or no one has appealed or made a complaint.

Information to help you evaluate assessment and IQA practice includes statistics such as enrolment, retention, success and progression data. This could affect the amount of funding received or future targets. Feedback from meetings and standardisation activities can also influence the way you evaluate various aspects. For example, you might decide to redesign some of the assessment documentation or the way that forms are completed, e.g. electronically rather than paper based.

Any issues or trends and areas of good or poor practice you have identified throughout the IQA process should be summarised and fed back to your assessors, or used as a standardisation activity.

Evaluation of both practice and systems can improve the service everyone receives and contribute to future planning within the organisation.

Surveys and questionnaires

A *survey* is a way of gathering information and data that could involve a wide variety of collection methods. For example, analysing previous data, carrying out observations and/or obtaining statistical information.

A *questionnaire* is a way of gathering information and data which usually involves oral or written questions given to an individual.

Both are useful ways of formally obtaining feedback from everyone involved in the training, assessment and IQA process. If you use surveys and questionnaires, you need to consider what you want to find out, who you will ask, when and why. Don't just ask questions for the sake of it, and don't just give them to those you know will give you a positive response. Everyone should be given the opportunity to be involved and then left to decide if they wish to respond or not.

When writing questions for a survey or questionnaire, you need to gauge the language and level to suit your respondents. You might be able to use jargon or complex terms with assessors, but not with others. The type of question is also crucial as to the amount of information you require. Using a *closed* question, i.e. a question only requiring a yes or no response, will not give you as much information as an *open* question, which enables the respondent to give a detailed answer.

Example

Did you receive a detailed assessment plan? YES/NO

Was the assessment activity as you expected? YES/NO

Did you receive feedback? YES/NO

Was your assessor supportive? YES/NO

These closed questions would not help you to understand what it was that the learner experienced, and they might just choose yes to be polite. It would, however, be easy to add up the number of yes and no responses to gain *quantitative data*. Think of this as the *quantity* of something, i.e. in terms of the total number of yes and no responses.

The questions would be better rephrased as open questions to encourage learners to answer in detail. This would give you *qualitative data*, therefore giving you more information to act on. Think of this as the *quality* of something which gives you information to work with rather than just data. A better way of wording the questions in the previous example would be:

Example

How detailed was your assessment plan?

What assessment activity was used and why?

How did you receive feedback?

How supportive was your assessor?

Using questions beginning with *who, what, when, where, why* and *how* (WWWWWH) will ensure you gain good quality answers. If you would rather use questions with yes/no responses, you could ask a further question to enable the learner to elaborate on why they answered yes or no.

Example

Was the assessment activity as you expected? YES/NO

Why was this?

This enables the learner to expand on their response, and gives you more information to act on. When designing questionnaires, use the KISS method: *Keep It Short and Simple*. Don't overcomplicate your questions, for example, by asking two unrelated questions in one sentence, or make the questionnaire so long that someone will not want to complete it.

You could consider using the Likert (1932) scale, which gives respondents several answers to a question, of which they need to choose one, such as:

1. Strongly disagree

2. Disagree

3. Neither agree nor disagree

4. Agree

5. Strongly agree

However, you might find respondents choose option 3 as a safe answer. Removing a middle response and giving four options forces a choice:

1. Strongly disagree

2. Disagree

3. Agree

4. Strongly agree

Anonymity for a survey or questionnaire might lead to you gaining more truthful responses if the person knows they will not be identified. However, if the respondent works closely with you, this might not be the case; the same goes for telephone, text message or face-to-face questioning. Electronic questionnaires that are e-mailed back will denote who the respondent is; however, postal ones will not (unless a reference code has been added to them). There are lots of online programs for surveys that will guarantee anonymity and will also analyse the results of quantitative data; some of these offer a free basic service such as www.surveymonkey.com. You could carry out a small sample survey with just a few learners to see how it works first. This is called a *pilot* and allows you to make any changes before a full survey is carried out, if necessary.

Activity

Design a short survey or questionnaire that could be used with learners or assessors. Consider the types of questions you will ask and how you will ask them, based upon the information you need to ascertain. Decide how the questionnaire will be implemented, e.g. paper based, online, in person. If possible, use it, analyse the results and consider what improvements could be made.

Always set a date for the return of any surveys and don't be disappointed if you don't get as many replies as you had hoped. Denscombe (2014) predicted a 30 per cent response rate, which isn't very high. However, if you give people time to complete a questionnaire, perhaps immediately after meeting them, they will hand it in straight away rather than take it away and forget about it. However, this will not ensure anonymity. Online questionnaires enable people to complete them in their own time, but a target date will need to be set.

Always inform your respondents why you are asking them to complete the questionnaire and what the information will be used for. Make sure you analyse the results, create an action plan and follow this through, otherwise the process is meaningless. Informing the respondents of the results and subsequent action keeps them up to date with developments and shows that you take their feedback seriously.

You could gain informal feedback from your staff after you have observed them or when you next meet with them. This will help you realise how effective you were and what you could improve in the future. It may also help you identify any problem areas, enabling you to do things differently next time. You could also encourage your assessors to gain informal feedback from their learners after they carry out an assessment activity, for example, after an observation in the workplace. You might also want to contact previous learners or employees who have left, to find out where they are now and how useful their training was to their career plan.

Always make sure you do something with the feedback you receive to help improve the product or service offered to everyone involved in the assessment and IQA process.

Self-evaluation

Self-evaluation is a good way of continually reflecting upon your own practice to ensure you are carrying out your role effectively. When evaluating your practice, you need to consider how your own behaviour has impacted upon others and what you could do to improve.

A straightforward method of reflection is to have an experience, then describe it, analyse it and revise it (EDAR). This method incorporates the WWWWWH approach and should help you consider ways of changing and/or improving.

Experience ⟶ Describe ⟶ Analyse ⟶ Revise (EDAR)

- **E**xperience – a significant event or incident you would like to change or improve.

- **D**escribe – aspects such as who was involved, what happened, when it happened and where it happened.

- **A**nalyse – consider the experience more deeply and ask yourself how it happened and why it happened.

- **R**evise – think about how you would do it differently if it happened again, and then try this out if you have the opportunity.

As a result, you might find your own skills improving, for example, giving more effective, constructive and developmental feedback to your assessors.

Reflection

Reflection should become a habit, for example, mentally running through the EDAR points after a significant event. As you become more experienced and analytical with reflective practice, you will progress from thoughts of *I didn't do that very well* to aspects of more significance such as *why* you didn't do it very well and *how* you could change something as a result. You may realise you need further training or support in some areas therefore partaking in relevant continuing professional development (CPD) should help.

There are various theories regarding reflection. Schön (1983) suggests two methods:

- reflection in action
- reflection on action.

Reflection *in action* happens at the time of the incident, is often unconscious and allows immediate changes to take place. It is about being *reactive* to a situation and dealing with it straight away.

Reflection *on action* takes place after the incident and is a more conscious process. This allows you time to think about the incident, consider a different approach or to talk to others about it before making changes. It is about being *proactive* and considering measures to prevent the situation happening again in the future.

Example

Aalia, the IQA, was observing Mark, an assessor, with a group of learners in a welding workshop. She noticed one of them was not wearing a visor correctly, which could lead to an accident. She immediately went over to the learner and showed him how to wear it the right way. This enabled her to deal with the situation at once. On reflection, she felt she should have asked Mark to deal with his learner. When giving Mark feedback after the observation, Aalia gave him additional advice on the importance of health and safety.

Continuing professional development

Continuing professional development (CPD) can be anything that you do that helps you improve your practice. It shows you are a committed professional and it should help improve your skills, knowledge and understanding. CPD should relate to your job role as well as the subjects you quality assure and can be based on the results of evaluation and feedback.

You should reflect on the activities you carry out so that they have a positive impact upon your development and role. There are constant changes in training and education; therefore it is crucial to keep up to date and embrace them. Examples include changes to the qualifications or standards you quality assure, changes to policies and practices within your organisation, regulatory requirements and government initiatives. It's useful to add relevant CPD activities to your curriculum vitae, particularly if you are applying for a job or a promotion.

CPD can be formal or informal, planned well in advance or be opportunistic, but it should have a real impact upon your role and lead to an improvement in your practice. CPD is more than just attending events or carrying out research; it is also about using reflection regarding your experiences which results in your improvement and/or development in your job role.

Your organisation might have a strategy for CPD, which will prioritise activities they consider are important to improving standards, and may or may not provide any funding for them. However, you can partake in lots of activities which are free or cost very little, for example, observing colleagues, researching websites, reading journals and textbooks.

Activity

Look at the following list and decide which of the activities would be relevant to you. What other activities could you carry out which would contribute towards your CPD?

Examples of opportunities for CPD include:

- attending events and training programmes
- attending meetings
- e-learning and online activities
- evaluating feedback from peers, assessors and others
- formally reflecting on experiences and documenting how it has improved practice
- improving own skills such as English, maths and ICT
- keeping up to date with relevant legislation
- membership of professional associations or committees
- observing colleagues
- reading textbooks and journals
- researching developments or changes to your subject
- secondments
- self-reflection
- shadowing colleagues
- standardisation activities
- studying for relevant qualifications
- subscribing to and reading relevant journals and websites
- using social media to follow/inform others of relevant and current information
- voluntary work
- work experience placements
- writing or reviewing books and articles.

It's a good idea to maintain a record of all CPD undertaken to prove you are remaining current within your role and your subject area. You could keep a manual record, such as the one shown in Table 2.12, or an electronic record, perhaps as a spreadsheet or by using specialist software.

Using a reference number for each activity enables you to cross-reference the activities to your documentation, for example, number 1 could be minutes of meetings, number 2 could be a certificate, number 3 a record of achievement (whether manual documents or electronic file names). Adding the reference number to the relevant documents also enables you to locate them when necessary. Besides keeping your CPD record up to date, you should write a more detailed reflection of what you learnt and how it impacted upon your job role.

Table 2.12 Example CPD record

Continuing professional development record

Name: Sue Smith   Organisation: The Acorn Company

Date	Activity and venue	Duration	Justification towards IQA role and subject area	Action required	Ref. no.
06 Jan	Attendance at an awarding organisation event. Heard about changes to the qualification and had a chance to discuss these with IQAs from other organisations.	2.5 hrs	Helped me understand the changes to the qualification to support my assessors.	Discuss the changes with the assessors at the next team meeting	1
10 Feb	Attendance at a First Aid training day.	6 hrs	To ensure I am current with First Aid in case someone has an accident.	–	2
20 Mar	Attendance at annual IQA training event. We discussed the types of records we use and how they could be improved, external reports and actions were analysed, and we updated the rationale and strategy for each subject area.	3 hrs	Ensured I am up to date with IQA practice	Disseminate relevant information to the assessors in my team.	3

You might participate in an appraisal or performance review system at some time, for example, with your manager. This is a valuable opportunity to discuss your progress, development and any training and/or support you may need. Having the support of your organisation will help you decide what is relevant to your development as an IQA, and towards your job role and subject area. An appraisal or an informal discussion is also a chance to reflect upon your achievements and successes. Always keep a copy of any documentation relating to your training and CPD, as you may need to provide this to funding bodies, awarding organisations or regulatory bodies if requested.

Keeping up to date

The following websites are useful to gain up-to-date information regarding developments in the further education and skills sector. Most of them enable you to register for electronic updates.

Department for Business, Innovation and Skills – www.bis.gov.uk

Department for Education – www.gov.uk/government/organisations/department-for-education

Education and Training Foundation – www.et-foundation.co.uk

Equality and Diversity Forum – www.edf.org.uk

FE News – www.fenews.co.uk

FE Week – www.feweek.co.uk

Government updates: Education and Learning – www.gov.uk/browse/education

National Institute of Adult Continuing Education – www.niace.org.uk

Ofqual – www.ofqual.gov.uk

Ofsted – www.ofsted.gov.uk

Teacher Educator UK – https://teachereducatoruk.wikispaces.com

Times Educational Supplement Online – www.tes.co.uk

Tutor Voices – www.facebook.com/groups/tutorvoices/

UKFEChat – www.ukfechat.com

Activity

Decide on a system for documenting your CPD if you don't already have one. You could use a form as in Table 2.12, or you could design your own. Add some recent activities to it and reflect upon each activity you have carried out, and how it has impacted upon your role as an IQA. If you have time, look at the websites listed in the previous bullet list and subscribe for updates from relevant ones.

You could join free social network sites such as LinkedIn, which is a professional networking site. Here you will find groups you can join specifically aimed at your specialist subject. You can post questions and respond to queries, and join in regular discussions.

You can follow me on LinkedIn, Twitter, Facebook and Google+ for the latest updates regarding what's happening in the further education and skills sector. Just search online for *Ann Gravells*. I also have a YouTube channel with several playlists of useful videos.

Following relevant people or organisations via social media will enable you to keep up to date with what's currently taking place.

Part of reflection is about knowing what you need to change. If you are not aware of something that needs changing, you will continue as you are until something serious occurs. Maintaining your CPD, keeping up to date with developments in your subject area, changes in legislation, changes in qualifications or standards and developments with ICT will assist your knowledge and practice.

Extension activity

Reflect upon a recent meeting you have chaired or attended. Evaluate how the meeting went, how you reacted to situations and what you could do differently next time. Consider what CPD you might need to help you with your internal quality assurer role.

Summary

Carrying out your role as an IQA in an organised and professional manner will help ensure assessment decisions are accurate, consistent, valid and reliable.

You might like to carry out further research by accessing the books and websites listed at the end of this chapter.

This chapter has covered the following topics:

- Internal quality assurance planning

- Internal quality assurance activities

- Making decisions

- Providing feedback to assessors

- Record keeping

- Evaluating practice

References and further information

Belbin, M. (2010) *Team Roles at Work* (2nd edition). Oxford: Elsevier Science & Technology.

Berne, E. (1973) *Games People Play: The Psychology of Human Relationships*. London: Penguin.

Denscombe, M. (2014) *The Good Research Guide*. Maidenhead: OU Press.

Goleman, D. (1995) *Emotional Intelligence*. London: Bloomsbury.

Likert, R. (1932) A Technique for the Measurement of Attitudes. *Archives of Psychology,* 140: 1–55.

Read, H. (2012) *The Best Quality Assurer's Guide*. Bideford: Read On Publications Ltd.

Roffey-Barentsen, J. and Malthouse, R. (2013) *Reflective Practice in Education and Training* (2nd edition). Exeter: Learning Matters.

Scales, P., Pickering, J., Senior, L., Headley, K., Garner, P. and Boulton, H. (2011) *Continuing Professional Development in the Lifelong Learning Sector*. Maidenhead: OU Press.

Schön, D. (1983) *The Reflective Practitioner*. San Francisco: Jossey-Bass.

Wood, J. and Dickinson, J. (2011) *Quality Assurance and Evaluation in the Lifelong Learning Sector*. Exeter: Learning Matters.

Websites

Association for Neuro Linguistic Programming – www.anlp.org

Belbin Team Roles – www.belbin.com

Data Protection Act 1998 – www.legislation.gov.uk/ukpga/1998/29/contents

Emotional Intelligence – www.unh.edu/emotional_intelligence/index.html

FE News – www.fenews.co.uk

Freedom of Information Act 2000 – www.legislation.gov.uk/ukpga/2000/36/contents

Ofqual – www.ofqual.gov.uk

Ofsted – www.ofsted.gov.uk

Online surveys – www.surveymonkey.com and www.smartsurvey.co.uk

Plagiarism – www.plagiarism.org and www.plagiarismadvice.org

Questionnaire design – www.wireuk.org/ten-steps-towards-designing-a-questionnaire.html

Self-evaluation – www2.warwick.ac.uk/services/ldc/resource/evaluation/tools/self/

quality assurance (IQA) activities taking place, for example, a large organisation offering training programmes which has several sites in different locations.

External quality assurance takes place for accredited qualifications and endorsed programmes which are delivered and assessed in an approved centre. An accredited qualification can be offered by several awarding organisations and is recognised nationally. Endorsed qualifications have usually been specifically written by a centre, in conjunction with one awarding organisation, to meet the needs of particular employers or learners.

Why carry out external quality assurance?

There needs to be a system of monitoring the performance and decisions of trainers, assessors and internal quality assurers within a centre. If not, staff might make incorrect judgements or pass a learner who hasn't met all the requirements, perhaps because they were biased towards them or had made a mistake.

As an external quality assurer, you will monitor a centre's activities for your particular subject area. You should remain objective and professional, and not become personally involved with the centre staff. Otherwise, this could affect your ability to make impartial decisions. You will need to inform the AO if you have worked in or know any of the staff in a centre you are allocated to, as this could be classed as a conflict of interest.

An effective external monitoring system will:

- ensure accuracy and consistency of training, assessment and internal quality assurance decisions

- uphold the credibility of the qualification or programme of learning, as well as the reputation of your own organisation and the awarding organisation, by ensuring compliance with all relevant requirements, regulations and standards

- support the centre staff by giving relevant up-to-date advice and guidance.

The EQA process is a quality mechanism to ensure the staff are maintaining compliance with the AO's requirements. The staff in the centre will include teachers, trainers, assessors and IQAs. Depending upon the type of qualification you will EQA, teaching and training may or may not be sampled. For example, some qualifications are formally delivered and require the use of schemes of work and session plans. Others might just be assessed, i.e. the competence of staff who are already performing the role in the workplace. You will need to find out if you should monitor training as well as assessment practice.

The EQA will want to ensure that all activities have been conducted in a consistent, safe and fair manner, for example:

- consistent: all staff are using similar training and assessment methods and making similar decisions across all learners. All learners have an equal chance of receiving an accurate decision

- safe: the methods used to train, assess and internally quality assure are ethical; there is little chance of plagiarism by learners; the work can be confirmed as authentic

confidentiality was taken into account; learning was not compromised, nor was the learner's experience or potential to achieve (safe in this context does not relate to health and safety but to the assessment and IQA methods used)

- fair: the methods used are appropriate to all learners at the required level and take into account any particular learner needs. Activities are fit for purpose, and planning; decisions and feedback are justifiable and equitable.

To become an external quality assurer, you may need to work for an awarding organisation; this could be on a part-time, or on a freelance or self-employed basis. It might even be on a zero-hours contract whereby work is not guaranteed. It's also possible that you might have to perform some tasks, for example, administration, in your own time.

The AO will often advertise any vacancies on their website. AOs prefer their external quality assurers to remain current with their practice in the areas they are quality assuring, therefore not many are employed full time. For example, if you are going to EQA a welding qualification, you might currently be working in a centre as a trainer and assessor of a welding programme. You should have considerable experience in the subjects you will externally quality assure and either hold or be working towards a recognised EQA qualification. If you haven't already done so, you should obtain a copy of the qualification specification or other criteria that you will externally quality assure and familiarise yourself with the content and requirements.

If you are currently working as a trainer, assessor and/or an internal quality assurer, this is not a conflict of interest with an EQA role, providing you declare where you are working to the AO. It's actually a benefit, as it means you are current with your knowledge and practice. Other terms are sometimes used for the EQA role, for example, external verifier, quality consultant, qualification consultant, and standards' verifier. The term EQA will be used throughout this chapter.

There are other forms of external quality assurance which you might come across, for example, external *moderation*. External moderators will sample a proportion of assessed work for a particular aspect. If the sample finds problems, then the work of every learner will need to be reviewed, including those outside of the original sample. External moderation is sometimes referred to as re-assessment to confirm the original decisions were accurate (or not).

Example

Liang is considering becoming an external quality assurer. He is currently working as an assessor and an internal quality assurer of carpentry and joinery qualifications. He has been looking at vacancies on several awarding organisations' websites and notices the job titles differ. Upon looking at the job descriptions, he sees they are all very similar. He notices there is a position available for a part-time EQA to cover the north of England, which is where he lives. He decides to complete the online application form as he thinks he fulfils all the criteria. He hopes that if he becomes an EQA, he will be able to identify areas of good practice which will also help him in his current role.

The external quality assurance cycle

Depending upon the subject you will externally quality assure, you will usually follow the EQA cycle. The cycle will continue to ensure the assessment and IQA process is constantly monitored. Records of all activities and decisions must be maintained throughout to satisfy the awarding organisation and the regulatory authorities. For example, in the UK there is Ofqual in England, DCELLS in Wales, CCEA in Northern Ireland and SQA in Scotland (links to these are at the end of the chapter in the website list).

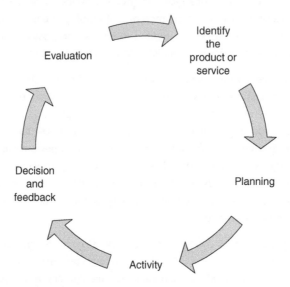

Figure 3.1 The external quality assurance cycle

The EQA cycle can involve the following aspects:

- **Identify the product or service** – ascertain what is to be externally quality assured and why. The criteria will need to be clear, i.e. units from a qualification (product) or the support the learner receives (service). External quality assurers might also carry out approval and advisory visits, as well as a monitoring visit or a remote monitoring activity to sample learners' work and records. These will be explained in Chapter 4.

- **Planning** – devise a sample plan to arrange what will be monitored, from whom and when. Plan the dates to observe staff and talk to learners and others involved such as witnesses. Information will need to be obtained from the centre staff (either by you or the AO) to assist the planning process, and risks taken into account such as staff knowledge and experience.

- **Activity** – various activities could be carried out, for example, a visit to a centre or a remote sample (where documents are posted to the external quality assurer or viewed electronically), to ensure quality and compliance of all aspects. Activities might also include approving centres to offer qualifications and making decisions to release certification rights (known as *direct claim status*). Centres which are consistent in their assessment and quality assurance approach will maintain their direct claim status. Awarding organisations will give advice regarding the activities that should be performed and when. Centre staff should comply with aspects in the qualification's assessment strategy, i.e. the qualifications and experience they should have. Concerns, issues, trends, complaints and appeals should always be identified and monitored.

- **Decision and feedback** – make a judgement as to whether the centre staff have performed satisfactorily, made valid and reliable decisions, and followed all required policies and procedures. Give developmental feedback as to what was good or what could be improved. Agree action and improvement points if necessary with appropriate target dates. (Action points are enforceable whereas improvement points are not.) Complete the awarding organisation's report and identify a risk rating of low, medium or high (see *risk ratings* later in this chapter).

- **Evaluation** – carry out a self-review of the EQA process to determine what is good practice, what could be improved or what could be done differently. Partake in the awarding organisation's review, appraisal and standardisation processes as required.

The cycle will then begin again with an identification of what needs to be monitored and when. Throughout the cycle, standardisation of practice between external quality assurers should take place; this will help ensure the consistency and fairness of decisions and the support given to centres. This might be at formal meetings, or via online webinars or another appropriate method. Records must be maintained of all activities. All external quality assurers should maintain their continuing professional development (CPD) to ensure their own practice is current and follow all relevant legal and organisational requirements. Please see Chapter 2 for some CPD ideas.

Extension activity

Look at the above bulleted list linked to Figure 3.1. Describe how each aspect will impact upon your role as an EQA. Is there anything you are unsure of or missing from the cycle? As you progress through Chapters 3 and 4, it should all become clearer.

Roles and responsibilities of an external quality assurer

Your main role will be to carry out the external quality assurance process according to the awarding organisation's requirements and follow all relevant regulatory and qualification guidelines. You must also ensure your centres are complying with relevant requirements, regulations and standards. You should advise and support your centres to ensure they maintain and improve the training, assessment and internal quality assurance processes. This should ensure that learners receive a good service and are able to reach their full potential. Learners will receive a certificate with the awarding organisation's name on it when they successfully complete. You therefore need to ensure you are maintaining the credibility of the qualification and the reputation of the AO as well as yourself.

EQAs should be organised, maintain relevant information, possess good time management and communication skills, and be able to deal with difficult situations should they occur. A sound knowledge and understanding of the subject area as well as of assessment and IQA systems is important for the role, particularly when making decisions regarding compliance. For example, if the subject area is motor vehicle maintenance, an EQA should not be externally quality assuring hairdressing. The process might be the same for each subject, but the external quality assurer must be fully familiar with the qualification content to make a valid and reliable decision and to support the centre. You will find it useful to read the current assessment and internal quality assurance guidance from your AO. This should be available on their website and will help you ensure your centres are meeting the requirements. Some qualifications and subjects state that you must also be a qualified assessor and/or internal quality assurer, as well as a qualified external quality assurer. This information will be in the qualification specification supplied by the AO.

Activity

If you are not currently working as an EQA, take a look at different awarding organisations' websites to see if they have any vacancies. Just search for 'EQA vacancies'. Have a look at the different roles available and see if you meet the requirements of their job descriptions. If you decide to apply for a position and are offered it, the AO might fund you to take a relevant EQA qualification.

If you are working as an EQA, find out from your AO if you can work towards an EQA qualification. If you've already started, take a look at the requirements to see how you can meet them.

You need to be aware that you will be representing the AO when communicating with centres. You should therefore act professionally and with integrity, remain impartial and objective, and maintain confidentiality where necessary.

Aspects of the external quality assurance role include (in alphabetical order):

- advising and supporting centre staff on an ongoing basis (not just during visits)
- communicating with centre staff and the awarding organisation on an ongoing basis
- completing a report of what was sampled, highlighting any action and improvement points, and judging whether the centre has a low, medium or high risk rating (action points are enforceable whereas improvement points are not, risk ratings will be covered later in this chapter)
- ensuring a centre's policies, procedures, systems and resources meet relevant requirements, regulations and standards
- ensuring centre staff interpret, understand and consistently apply the requirements of the qualification or programme of learning

- ensuring learners are registered with the awarding organisation within the required timescale (learners who are not registered should not be sampled)

- ensuring quality throughout the learner journey within a centre

- ensuring centre staff meet any requirements, regulations and standards, and standardise their practice

- ensuring the accuracy and consistency of trainer, assessor and internal quality assurer decisions

- evaluating and approving centres to offer qualifications, releasing certification rights when they are performing satisfactorily or recommending removal of direct claim status if necessary

- giving guidance to centre staff regarding the qualification criteria, relevant requirements, regulations and standards

- identifying issues and trends, for example, if all assessors are misinterpreting the same aspect or making the same mistakes

- keeping full and accurate confidential records

- monitoring and auditing the full learner journey from commencement to completion, e.g. information, advice and guidance (IAG), recruitment, initial assessment, induction, training, formative and summative assessment, decision making, feedback, support for progression opportunities

- monitoring risk within a centre, e.g. when new qualifications are introduced or if there is a high staff turnover

- observing training, assessment, feedback and IQA practice

- planning what will be monitored and communicating this to all concerned within the awarding organisation's timescales

- sampling assessed and internally quality assured learners' work (and records) according to a planned strategy, and making decisions based on facts

- training staff and giving support and advice, for example, as an additional visit to a centre if required

- updating own CPD regarding subject knowledge and EQA practice.

As you become more experienced, you might be asked to mentor and support new external quality assurers. This might involve shadowing them during a centre visit and giving advice and guidance as necessary. Hopefully, this type of support will be given to you when you first start, as the role can often feel quite isolated. If you are working towards an EQA qualification, you might also be shadowed and supported by another qualified external quality assurer. If you are responsible for a team of EQAs, you will find the content of Chapter 5 helpful.

| **Example** |

Mary is a new EQA and has been working as an assessor and an IQA in a centre for several years. She is due to be accompanied by an experienced EQA in the same subject area during a centre visit next month. Prior to this, she has a list of questions which includes: where to obtain the current documents for preparing the centre visit, how to ensure she is using the most up-to-date qualification standards, and can she see an example of a completed EQA report? Being prepared and obtaining all the relevant information and documentation will help Mary with her role.

Documents to perform your role can usually be located on the AO's website, for example, the qualification specification, assessment and IQA guidance, appeals and complaints procedures, and templates and reports. You might need to provide your own equipment and stationery but do check if any of this is provided by your AO. You will also need to find out who can give you support and advice, and whether you will be accompanied on your first centre visit.

Legislation, regulations, codes of practice, policies and procedures

These may differ depending upon the context and environment within which you will externally quality assure. You also might need to be aware of the requirements of external bodies and regulators such as Ofsted (in England) who inspect funded provision.

Legislation

There is a variety of relevant legislation which you may need to follow, such as The Data Protection Act 1998 and The Health and Safety at Work etc. Act 1974 which may impact upon your role. Please see Chapter 1 for further information regarding relevant legislation.

Regulations

Regulations are often called *rules* and they specify mandatory requirements that must be met. Public bodies, corporations, agencies and organisations create regulatory requirements which must be followed if they are applicable to your job role. For example, in education one of the regulators is Ofqual which regulates qualifications, examinations and assessments in England. Ofqual approves and regulates awarding organisations. There will also be specific regulations which relate to your specialist subject and you will need to find out what these are.

You will need to follow the awarding organisation's regulations as well as relevant regulations such as Ofqual's *General Conditions of Recognition* (2015). This document sets out certain conditions which awarding organisations must ensure their centres adhere to. Examples include managing conflicts of interest, identifying and managing risk, and dealing with malpractice.

Activity

Obtain a copy of the Ofqual (2015) General Conditions of Recognition for England (just type those words into an internet search engine) and have a look at the conditions. If you work outside England, find out which regulations will be applicable to you. Make a note of how they will impact upon your role as an external quality assurer when supporting a centre. If they have since been updated, make sure you are looking at the correct version.

Please see Chapter 1 for further information regarding relevant regulations, for example, the Control of Substances Hazardous to Health (COSHH) Regulations 2002 which applies if you visit centres which work with hazardous materials.

Codes of practice

Codes of practice are usually produced by organisations, associations and professional bodies. They can be mandatory or voluntary and you will need to find out which are applicable to you. For example, you might be required to interview young and/or vulnerable learners when visiting a centre and your AO should give you guidance regarding this. You could be required to undertake a criminal records check through the Disclosure and Barring Service (DBS).

Examples of codes of practice include:

- acceptable use of information and communication technology (ICT)
- code of conduct
- confidentiality of information
- conflict of interest
- disciplinary
- duty to prevent radicalisation
- environmental awareness
- lone working
- management of information and records
- misconduct
- sustainability.

Policies and procedures

There will be several policies and procedures issued by your awarding organisation with which you should become familiar, for example:

- access and fair assessment

- appeals and complaints

- bribery

- confidentiality of information

- copyright and data protection

- equality and diversity

- external quality assurance

- health, safety and welfare (including Safeguarding)

- malpractice.

Think of the policy as a statement of intent and the procedure as how the policy will be put into action. Policies don't need to be long or complicated, but should provide a set of principles to help with decision making. Procedures should state who will do what and when, and what documentation or checklists should be used.

Extension activity

Find out what legislation, regulatory requirements, codes of practice, policies and procedures relate to your role as an external quality assurer for your particular subject. Which are the most important and why? How will they impact upon your role and the external quality assurance process?

Concepts and principles of external quality assurance

Concepts relate to ideas, whereas principles are how the ideas are put into practice. For the purpose of this chapter, they have been separated for clarity; however, some concepts could also be classed as principles depending upon your interpretation.

Key concepts

Concepts are the aspects involved throughout the EQA process. They include:

- Competence – ensuring your own competence, skills and knowledge is up to date, not only regarding the job role, but also regarding subject knowledge of the qualification and standards being quality assured.

- Communication – ensuring this takes place regularly with centre and awarding organisation staff.

- Equality and diversity – ensuring all activities embrace equality, inclusivity and diversity, represent all aspects of society and meet the requirements of the Equality Act 2010.

- Ethics – ensuring all activities are honest and moral, and take into account confidentiality, integrity and transparency.

- Health and safety – ensuring these are taken into account throughout the full monitoring process, carrying out risk assessments as necessary and ensuring appropriate safeguarding procedures are in place.

- Record keeping – ensuring accurate records are safely maintained throughout the monitoring process. Records should also remain secure and confidential, and only be shown to those with a legitimate interest.

- Risk ratings – identifying whether a centre is low, medium or high risk – the awarding organisation will give guidance towards this, and it is covered later in this chapter.

- Standardisation – carrying out activities with other external quality assurers to ensure the requirements of the qualification or programme of learning are interpreted accurately, that they are making comparable decisions and giving consistent support to centres.

- Strategy – ensuring a written strategy is in place which clearly explains the full process of what will be externally quality assured, when and how, with each centre.

Key principles

Principles are based upon the concepts and relate to *how* the EQA process is put into practice.

One important principle is known by the acronym VACSR and should be followed when carrying out external quality assurance activities.

- **V**alid – assessor and IQA decisions and feedback are relevant to what has been assessed.

- **A**uthentic – the work has been produced solely by the learner and the assessor has confirmed this.

- **C**urrent – the learner's work is relevant at the time of assessment, as are the assessment and IQA records which have been used.

- **S**ufficient – the learner's work covers all of the requirements, and the assessor and IQA records are complete, legible and accurate.

- **R**eliable – the assessment and IQA processes are consistent over time and at the required level.

If the above are not checked, you will not be supporting your centres correctly. This could lead them to think their practice is acceptable when in reality it might not be.

Another key principle for an EQA is known by the acronym SMART – ensuring all activities are **s**pecific, **m**easurable, **a**chievable, **r**elevant and **t**ime bound.

Activity

Look at the previous bulleted lists of concepts and principles, and list those which you feel are relevant to your role as an EQA. Explain how you will perform each one in a way that supports your centres, whilst maintaining compliance.

Following the key concepts and principles of external quality assurance will ensure you are performing your role according to the relevant regulations and requirements.

The EQA rationale

A good EQA system will start with a *rationale*. This is the reason *why* EQA takes place and ensures the activities used are valid and reliable.

- Valid: the methods used are based on the requirements of the qualification or standards being sampled.
- Reliable: a similar decision would be made with similar staff from similar centres.

A rationale is usually produced by the awarding organisation to help maintain the credibility of the qualification and their reputation.

Example

All external quality assurance activities will be carried out by qualified and experienced staff. This is to support centres, maintain compliance and uphold the credibility of the qualification and the reputation of the awarding organisation. External quality assurers will monitor centre activity to ensure the safety, fairness, validity and reliability of assessment and internal quality assurance decisions throughout the learner journey.

EQA strategy

You might find it useful to create an EQA strategy for each centre you will monitor. This will give a structure to your planning and sampling activities. Your awarding organisation might have an EQA handbook which will have lots of useful advice for you to follow. If you don't already have a copy, make sure you ask where you can obtain one. Your strategy should be based on any possible risk factors identified within the centre and may change over time depending upon how the centre is performing.

Example

EQA strategy for a centre which offers the Level 2 Customer Service qualification (one internal quality assurer, four assessors, average of 100 learners). The EQA has been allocated to the centre for the next three years.

The external quality assurer will:

- *follow up previous EQA action points on an ongoing basis*

- *observe feedback from the internal quality assurer to each assessor over time*

- *observe the practice of each assessor over time*

- *interview a sample of learners and witnesses (in person or via the telephone)*

- *sample all levels, all units and all methods of assessment over time*

- *sample the planning, decisions and feedback records from all assessors over time, along with those of the internal quality assurer*

- *sample supporting materials such as initial assessment, induction, tutorial reviews, minutes of meetings, standardisation records, policies and procedures, evaluations with supporting data analysis, and appeals and complaints, over time*

- *maintain full records of all EQA activities and communications with the centre and AO.*

Factors to consider when creating a strategy

When planning your strategy, use the five W and one H model (WWWWWH) of who, what, when, where, why and how. This will help you consider the factors you need to take into account when planning your activities with a centre, i.e.

- **W**ho will you need to meet, observe or interview during the visit? Who will you need to communicate with during a remote sample?

- **W**hat activities will be carried out and what records will you need to complete?

- **W**hen will the activities take place?

- **W**here will the activities take place?

- **W**hy are you doing certain activities, for example, based on risk factors?

- **H**ow will you carry out the activities, and how will you make your decisions, complete your report and give feedback to centre staff?

Example

Andrea has just been allocated a centre as the previous external quality assurer has now retired. She has obtained the centre contact details from the AO and introduced herself by telephone. She has read the last three reports which did not contain any action points. As this is a small, low-risk centre, she has agreed

(Continued)

with the AO that she will visit in four months' time. Prior to the visit, she will obtain details of all relevant staff and learners to help her decide on her strategy of what to sample, from whom and when.

When planning your strategy, you should take into account factors such as (in alphabetical order):

- assessment methods: are they robust, safe, valid, fair and reliable? Are they complex and varied? Do they include online assessments, holistic assessment and/or witnesses testimonies?

- availability of assessors and IQAs

- experience, workload and caseload of assessors and IQAs (consider staff turnover)

- number of assessors to internal quality assurers (should be fairly allocated)

- number of learners to assessors (allocating learners to assessors should be fair, assessors should not be overloaded)

- problem units or aspects which could be misunderstood

- qualification content (staff might interpret aspects differently)

- types of records to be completed (manual or electronic).

You might also like to view the centre's most recent Ofsted report, if applicable, to get a feel for how they operate. These are freely available online and can be accessed by anyone; the web address is at the end of this chapter.

You might be alerted to a risk within a centre due to (in alphabetical order):

- a long period of no communication from the centre

- a whistle-blower or staff member informing you of something wrong (don't take their word for it, but pass it on to the AO to deal with)

- changes of staff or resources not communicated to you until you visit

- documents from the centre not being correctly completed or showing serious anomalies

- far too many or too few learner registrations and certifications

- inconsistencies with the information supplied by the centre compared to those the awarding organisation holds

- the awarding organisation informing you of an issue

- the look of the centre's website, i.e. spelling and grammar errors, dubious claims, the use of stock photos, a *contact us* form but no telephone number or address

- the requested information from the centre not being received by the due date, or being incomplete.

These are all aspects you will need to take into consideration to ensure the centre is functioning as it should.

Risk management

Whenever possible, it's best to be proactive regarding any risks a centre might pose, rather than being reactive to a situation after the event. If you are in contact with your centres on a regular basis, you could encourage them to inform you of any concerns they might have, in order for you to give relevant advice and support. For example, a learner might have been caught plagiarising another learner's work. You would therefore need to check that the centre had an adequate plagiarism policy and that it had been followed correctly. If you are not aware of a situation which might pose a risk, it could become quite serious by the time you visit the centre.

It's important to monitor and manage the risks a centre might pose to ensure adequate support can be given and that the learners are not disadvantaged in any way.

There are many *risk factors* to take into consideration when planning EQA activities, for example (in alphabetical order):

- achievement (or not) of previous action points
- appeals and complaints
- assessment methods used and types of evidence provided by learners

- assessor expertise, knowledge and competence, whether new, experienced, qualified or working towards an assessor/IQA qualification (staff should have appropriate job descriptions and partake in CPD)

- assessors (or teachers/trainers) who assess the same subject but with different groups of learners

- authenticity of learners' work

- case loads and pressures of work placed upon staff, for example, expected targets to be met, funding based on achievements, staff having unclear roles

- changes to qualifications, standards, documents, policies and procedures, and records

- language barriers

- locations of learners, assessors and internal quality assurers

- numbers of learners and how quick (or how long) they take to complete

- possible plagiarism by learners

- previous risk rating of the centre

- reliability of witnesses, if used, and how they are supported

- turnover of staff

- type of qualification or programme being assessed, problem areas or units

- use of appropriate holistic assessments and recognition of prior learning

- use of technology and its reliability in assessment and IQA

- whether evidence and records are manual or electronic

- whether the learners have been registered with the awarding organisation (an external quality assurer should not sample from a learner who is not registered).

You should also consider any risks regarding your own role, e.g. making an invalid decision, or giving inappropriate feedback because you are not up to date with the qualification or AO's requirements. You need to make sure your own practice is current and that your judgements are valid and reliable.

Risk ratings

Most awarding organisations grade their centres on a risk rating of low, medium or high. However, a new centre, one with new/inexperienced staff or very large numbers of learners might mean it is classed as high risk even though nothing has been done wrong. An EQA therefore needs to build up confidence in the centre and carry out several monitoring activities to ensure the centre is operating correctly. Over time, the more serious a risk a centre poses, the more serious a sanction they will be given. A sanction can mean a centre losing their direct claim status, i.e. they can't apply for certificates unless an external sample takes place. More serious sanctions include a centre losing both registration and certification rights until they put right all the problems. If you have concerns, it's best to talk to the awarding organisation before completing your report and giving feedback to the centre.

Awarding organisations will give guidance as to how to make a decision which results in a low, medium or high risk rating being attributed to a centre (see Table 3.1). Although you can recommend a particular risk rating for a centre, it is the AO who makes the final decision. If you feel a centre should be placed on a higher risk rating during a visit, you could discreetly leave the room to make a phone call to the AO to discuss your findings. As the final decision rests with them regarding risk ratings and sanctions, you must contact them first. If you can't get in touch, you can make a recommendation on your report, but you must inform the centre that the AO makes the final decision.

Placing a centre on sanctions or changing their status to a high risk rating can affect their operation and any funding they might receive. You might feel pressured by staff during a centre visit not to place them on a higher risk rating because of this. You must always remain objective and ensure you do everything *by the book*, in accordance with relevant guidelines and regulations. If you are in doubt about anything, give your AO a call.

At the end of your monitoring activity, you must make a valid and reliable decision as to whether the centre can continue to operate as low risk or if you need to recommend a higher risk rating to the AO. You can give action points and target dates for any minor issues which are a low risk, however, these should always be followed up to ensure they have been met.

Table 3.1 Examples of risk ratings (these may differ depending upon the AO)

Risk rating	Meaning	Sanction
Low	The centre is complying with all awarding organisation, qualification and regulatory requirements. There might be some minor issues that need addressing therefore an appropriate action plan can be agreed with the centre.	The centre is not sanctioned and can register and certificate their learners.
Medium	There is some non-compliance, for example, insufficient IQA record keeping.	The centre can register their learners, but cannot claim certificates unless the external quality assurer carries out a further sample.
High	There are serious non-compliance issues e.g. insufficient staff, no internal quality assurance system in place, inaccurate data maintained. Or the centre is new, staff are inexperienced or there is a high volume of learners.	The centre cannot register any more learners or claim certificates for existing learners. High risk can also mean a centre having their approval withdrawn from offering a particular qualification if aspects are very serious.

Malpractice

Malpractice can include learners plagiarising each other's work, assessors signing off units which are incomplete and internal quality assurers completing records for aspects they have not sampled. It could be intentional or accidental and you would tactfully need to find out which.

Activity

If you are currently working as an assessor, an internal or external quality assurer, obtain a copy of the awarding organisation's quality manual and find out what their risk ratings mean, what sanctions can be imposed and what their malpractice procedure is.

If you suspect malpractice, don't bring this to the attention of the centre staff: contact your AO immediately and discuss how to approach it with them first. They might ask you to sample the current learners who have completed the qualification to allow certificates to be claimed. However, future learners' certificates cannot be claimed until the required action points have been met.

Other areas where malpractice could occur are (in alphabetical order):

- an internal quality assurer overruling an assessor (due to pressures to meet targets) when the assessor did not pass the learner

- assessment records being completed and signed when assessments did not take place

- certificates being claimed for learners who do not exist or who have not yet completed

- dates of commencement and achievement not agreeing with those that the learners tell you

- dates on centre records not matching those when the activities took place

- learners' work and supporting records which you have requested are not available, belong to someone else or have missing items

- minutes of meetings being produced when they didn't actually take place

- signatures on documents not matching those of the people concerned

- standardisation records being completed for activities that did not take place.

It will be the responsibility of the awarding organisation, not yourself, to follow up any serious issues. You should not inform your centre that an investigation might take place as this information should remain confidential. If an investigation is carried out, you may be asked to give a statement.

What will you need to do when your awarding organisation allocates a centre to you? Make a list of the information you will need to obtain from both the awarding organisation and the centre. Consider how you will use this information to plan your first visit monitoring activity with a centre. What risks might occur which could lead to malpractice within a centre, for your specialist subject? How would you deal with it?

The role of technology in external quality assurance

Information and communication technology (ICT) can be used to support and enhance the EQA process, as well as being used within a centre to support learning and assessment. Technology is constantly evolving and new resources are frequently becoming available. It's crucial to keep up to date with new developments and you should try to use these whenever possible. Communication through e-mail or web-based platforms can simplify the contact process between yourself and the centre staff, as well as with the AO staff.

Online assessment (or electronic assessment: e-assessment) might be used in a centre in addition to other assessment methods. For example, learners could complete a multiple choice test which will automatically be marked and give instant results. The system could generate different questions for each learner so that no two tests are the same. Tests can be taken on demand when a learner is ready and can be taken anywhere there is a suitable device with access to the software. However, if supervision is necessary for online tests, you will need to ensure this aspect is quality assured at the centre.

Activity

Consider what aspects of the EQA role could be carried out using technology, either during a visit or a remote sample, or between visits to a centre. You don't need to be an expert at using the different systems centres might use. The centre staff should support you with access. However, don't feel obliged to see only what they want you to see.

Examples of using ICT for the EQA role include (in alphabetical order):

- accessing an electronic sample of learner work, assessment and IQA records remotely online rather than visiting the centre

- accessing the awarding organisation's website to locate documentation

- communicating via e-mail, Skype, or particular website platforms, forums or networking sites

learners enrolled, or just a few. You will not be expected to know all the different systems available; therefore you can ask your centre to explain to you how they use theirs.

Example

Pedro had just been allocated a new centre. Upon his initial contact with the IQA he ascertained the centre used an e-portfolio system. He identified the main issues to be ensuring the authenticity of learners' work, accessing accurate assessment and IQA records, and the tracking of learners' progress and achievements.

When he agreed the visit date, he asked that the centre have a member of staff available to demonstrate the system to him. He also asked that this person remain contactable throughout the visit, in case he encountered any problems accessing anything.

As an EQA, it's useful to be aware of some of the issues regarding online assessment, for example (in alphabetical order):

- availability and access to the software or applications being used, familiarity with ICT, internet availability and download speeds
- encouraging learners to update their profiles, agreeing a social etiquette and communicating with others (e.g. assessors and those within the online community)
- ensuring learners are fully aware of what they need to do, along with target dates
- ensuring the authenticity, safety and security of the learners' work and data
- how the assessment and IQA records are maintained and how they can be accessed by others
- knowing when to intervene or moderate in a situation that might become out of hand, for example, issues not related to the subject
- learners using their own devices which might enable them to gain access to the answers to any assessment questions, or communicate with other learners during a test
- motivating learners to establish a routine, for example, to commit regular times for study, online communication and discussions, and the submission of work
- that screen sizes differ therefore some people might struggle to read on a small screen
- the support each individual will require and what they actually receive
- time management for the planning of synchronous learning, online assessment activities and communication to take place in real time.

Being aware of some of the issues should help you plan what questions you will need to ask the centre staff.

If you are currently working as an EQA, identify the issues which could occur when centres use technology for training and assessment. List any concerns you might have. You may need to discuss these with your awarding organisation if you are unsure how to deal with them.

Summary

Following the key concepts and principles of external quality assurance will ensure you are performing your role as an EQA according to all the relevant requirements, regulations and standards. It will also help you to support your centres' staff.

You might like to carry out further research by accessing the books and websites listed at the end of this chapter.

This chapter has covered the following topics:

- External quality assurance

- Roles and responsibilities of an external quality assurer

- Concepts and principles of external quality assurance

- Risk management

- The role of technology in external quality assurance

References and further information

Ofqual (2015) *General Conditions of Recognition*. Coventry: Ofqual.

Pontin, K. (2012) *Practical Guide to Quality Assurance*. London: City & Guilds.

Read, H. (2012) *The Best Quality Assurer's Guide*. Bideford: Read On Publications Ltd.

Wilson, L. (2012) *Practical Teaching: A Guide to Assessment and Quality Assurance*. Hampshire: Cengage Learning.

Wood, J. and Dickinson, J. (2011) *Quality Assurance and Evaluation in the Lifelong Learning Sector*. Exeter: Learning Matters.

Websites

Council for the Curriculum, Examinations and Assessment in Northern Ireland (CCEA) – http://ccea.org.uk

Data Protection Act 1998 – www.legislation.gov.uk/ukpga/1998/29/contents

The Department for Children, Education, Lifelong Learning and Skills in Wales (DCELLS) – http://gov.wales/topics/educationandskills/?lang=en

Disclosure & Barring Service – www.gov.uk/government/organisations/disclosure-and-barring-service

Equality Act 2010 – www.homeoffice.gov.uk/equalities/equality-act/

Federation of Awarding Bodies – www.awarding.org.uk

Health and Safety at Work etc Act 1974 – www.hse.gov.uk/legislation/hswa.htm

Ofqual – www.ofqual.gov.uk

Ofqual General Conditions of Recognition – www.gov.uk/government/uploads/system/uploads/ attachment_data/file/461218/general-conditions-of-recognition-september-2015.pdf

Ofsted inspection reports – www.ofsted.gov.uk/inspection-reports/find-inspection-report

Plagiarism – www.plagiarism.org and www.plagiarismadvice.org

Scottish Qualifications Authority (SQA) – www.sqa.org.uk

4 PRACTICES OF EXTERNAL QUALITY ASSURANCE

Introduction

External quality assurance activities will help ensure a centre's assessment and internal quality assurance decisions are accurate, consistent, valid and reliable. As an external quality assurer, you will be representing an awarding organisation and will monitor a qualification in your specialist subject area. You will therefore need to safeguard the credibility of the qualification and the reputation of the awarding organisation. You will need to make sure the centres are complying with all the relevant requirements, regulations and standards, as well as giving guidance and support as necessary.

Your role might just be a small part of your profession, for example, if it's carried out on a part-time or a freelance basis. You might also be an assessor and/or an internal quality assurer in a centre, which is a good way of maintaining your currency of practice.

This chapter will explore how you can plan and carry out various external quality assurance activities with centres.

This chapter will cover the following topics:

- External quality assurance planning
- External quality assurance activities
- Making decisions
- Providing feedback to centre staff
- Record keeping
- Evaluating practice

External quality assurance planning

When appointed as an external quality assurer (EQA) for an awarding organisation (AO) you will usually be allocated to centres to enable you to start planning your quality assurance activities. It's advisable to be fully familiar with assessment and internal quality assurance (IQA) approaches for the subject you will externally quality assure. This will help you support your centres and know what you are looking for when sampling. The number of centres you will be allocated and their locations will depend upon the time you have available to devote to the role. Not all EQAs are employed full time. Many are part time and still work as an assessor or an internal quality assurer (IQA) in a centre. When you start, you should be given a job description and advice, support and training from the AO. If you don't have one, then working towards the requirements of an appropriate external quality assurance (EQA) qualification or following national occupational standards will ensure you are performing your role effectively.

If you are working for more than one AO, you might find things will differ between them. You will need to be careful not to get mixed up with what you are required to do and what documents you need to use. You should inform the AO of any other employment you have to ensure there isn't a conflict of interest. For example, if you work as an assessor or an IQA in a centre, you will not be allowed to be an EQA for the same centre.

If you are not employed by an AO, but are carrying out a role similar to that of an EQA within your own organisation, the content of this chapter will still apply, for example, if you are externally quality assuring in-house training programmes which are assessed and internally quality assured (IQA) across various sites and locations.

If you have been externally quality assuring for a while, you may find your AO reallocates your centres to another EQA after a period of time. This is to ensure you and other external quality assurers are not becoming too complacent or too familiar with how a particular centre operates. Another external quality assurer could see things that you might not.

Example

Karina had been an external quality assurer for two centres for over three years. She had got to know the centre staff very well and was familiar with their policies and procedures. Her reports were very favourable and there were never any action points or sanctions. Although Karina had done nothing wrong, the awarding organisation now felt it was time for a change of EQA. This would enable someone else to view things with a fresh look. Karina was therefore allocated to two other centres.

Some AOs allocate their EQAs to different centres all the time, rather than to particular centres for a set amount of time. Whilst this does not enable a working relationship to be built up over time, it does allow for impartiality and objective decision making. In these cases, the AO often carries out the communication with the centre and allocates an available EQA when necessary.

All centres have a *centre number* which will be given to you by the AO, and you should check the centre uses the same number on all their documentation. Once allocated to a centre, if this is part of your role, you will need to make contact and obtain certain information about the staff and the qualifications they offer. This might include:

- communication details: phone number and e-mail address of the key contact person, often called a centre co-ordinator or head of centre
- address, location and details of different assessment sites

- details of staff involved: names, curriculum vitaes, qualifications and experience

- details of centre policies and procedures, e.g. appeals, complaints, assessment, IQA

- details of qualifications offered and levels, dates approved, date certification was released

- numbers of learners and how they are allocated to assessors.

Some or all of this information might be given to you by the AO or the previous EQA. However, you will still need to contact the centre to introduce yourself and let them know when you will carry out a relevant EQA activity. The named contact person might be an administrator and not the person who is able to give you all the specific information you need. You will therefore need to find out who it is at the centre you need to get in touch with.

Activity

What information, besides that in the previous bulleted list, might you need to obtain from the centre and/or the awarding organisation (or previous EQA) to help you perform your role?

You should obtain and read any previous EQA reports and action plans regarding the centre to familiarise yourself with how they operate. You should also obtain and read copies of the awarding organisation's guidelines, the qualification specification for the subject you will externally quality assure (and the final dates for registration and certification) and any other relevant guidelines and regulations which relate to the subject being monitored.

You will also need to plan for using public transport if you are visiting a centre at a distance or find out a suitable driving route and car parking arrangements. You might need to provide your own refreshments and meals, or ask if the centre has facilities for you to access. Sometimes a centre will be happy to provide you with refreshments, a meal or a sandwich. However, you should check with your AO in case you can't accept anything due to their bribery policy.

Resources

You will need to have appropriate resources to carry out your role. Not all awarding organisations supply these for you and therefore you might need to provide some yourself.

Resources could include:

- appropriate clothing and name badge/identification card

- awarding organisation's documents, qualification specification, relevant regulations, reports and checklists (manual or electronic)

- briefcase, pens, paper

- computer/laptop/tablet or other device with internet access

- insurance, such as professional indemnity and business car insurance (if you are not covered by the AO)

- money for expenses, e.g. bridge tolls, meals, transport costs (some of which you might be able to reclaim)

- telephone: landline and/or mobile.

Example

Andy is a new external quality assurer who has just been allocated five centres on a self-employed basis. This means he will only be paid when he carries out a monitoring activity, and he must declare his income to the Inland Revenue. However, he will be able to claim travel and subsistence expenses. He has received training from the awarding organisation and has been given electronic access to all the guidelines and documentation he needs. He has also been given a name badge and a pen with the AO logo on. He was surprised to find that he needed to provide everything else himself. This includes having professional indemnity and business car insurance, his own laptop, printer and telephone.

Recording ongoing contact with centres

When communicating with centres, it's useful to keep a record of contact to enable you to keep track of your activities with each individual centre. You could do this manually or electronically and an example is given in Table 4.1 on page 112. It can be more than a record of contact, as you can add details of the staff you have met, when you have checked their curriculum vitaes (CV) and when you have authorised the release of certification for particular qualifications and levels. If you communicate regularly via e-mail, you might like to create a folder for each centre and move the e-mails into it for ease of access.

Activity

Create a record of contact for each of the centres you have been allocated, similar to the one in Table 4.1 on page 112. Start completing them with information regarding your centres and staff. If they are electronic, create a folder on your computer to save all documents for each centre. Make sure you have a backup copy, perhaps on a USB stick or saved to a cloud drive. You will need to make sure you follow data protection and confidentiality requirements.

To perform your role fully, you will need to carry out certain administrative duties and use various documents, templates, checklists and reports. These will probably be supplied by the AO (electronic or manual) or you might need to create your own.

Table 4.1 Example centre record of contact

Centre details			
Centre name	XYZ Company	**Centre number**	0123456
Contact name	Joan Haverham	**Contact's position**	Internal quality assurer
Address	Manor Building, Main Road, MillerVille		
E-mail address of contact	j.haverham@xyzcompany		
Tel numbers Landline	01112 345678	**Mobile**	07123 456789

Qualification details				
Qualifications	Levels	AO ref no	Date approved	Date certification released
Customer Service - Certificate	1, 2	0123 01 and 02	03.04	05.09
Customer Service - Diploma	3, 4	0123 03 and 04	03.04	15.03

Staff details			
Name	Position	Met on	CV approved?
Jean Haverham	Internal quality assurer and centre co-ordinator (qualified assessor and IQA)	03.04	Y
Phil Scholey	Trainer/Assessor (qualified assessor)	03.04	Y
Terri Mandalay	Trainer/Assessor (working towards an assessor qualification)	03.04	Y

Contact log		
Date	Mode of contact*	Details
19.03	T	Discussed qualification application and arranged an approval visit date.
03.04	V	Met staff and completed qualification approval documents. E-mailed documents to awarding organisation and centre.
15.06	T	Rang Jean to check on progress of learners – no problems so far – 10 registrations on each level of the Certificate.
09.07	E	E-mailed 'AO update' to Jean, asked her to forward to other staff and to discuss at next meeting.
13.08	T	Jean rang to arrange a visit as most units were complete. Checked with AO for approval to plan a visit.
05.09	V	EQA visit – see completed documents and report to support release of certification for 0123 (01, 02). 0123 (03 and 04) have not been completed yet. See report for action points.
15.03	R	Remote monitoring activity of 0123 (03 and 04) - see completed documents and report to support release of certification. No action points but a few improvement points for the centre to consider.

*T – telephone call, E – e-mail, F – fax, L – letter, R – remote monitoring activity, V – visit

Planning what to sample

When you have agreed a date to sample a centre's activity, you can prepare a sampling strategy based on any identified risks. You might plan to carry out a remote monitoring activity (where the centre posts learner work and supporting records to you, or gives you electronic access to them) or you might plan to visit in person. If it's the latter, you will have a visit plan which will also incorporate a sample plan. Your AO will give you guidance regarding which type of activity is required and the timescales for requesting information and sending your plan to the centre.

Creating a sample plan will give a focus to the activities you wish to carry out with a centre. The plan should be sent to the centre in advance to inform the centre staff what you will do and when. It should be based on your EQA strategy which will be different for each centre according to any identified risks (see Chapter 3). You will need to ascertain certain information in advance to help inform what you will sample, for example:

- learner details, i.e. names, start dates, registration dates, unit completion dates, and final completion and certification dates (if applicable). You should be able to check this information with the AO records to ensure the centre is not missing anyone off for any reason. If they have, you will need to tactfully check if it was a mistake or deliberate

- trainer, assessor and IQA names, CVs and CPD details (depending upon when you last saw them). Details of all staff should be given to you, even if some have recently left, as they may have been involved with learners prior to leaving

- assessment dates, i.e. a copy of the assessment tracking sheet for each assessor to show you what has been assessed and when, for each learner

- IQA dates, i.e. a copy of the IQA sample plan and tracking sheet showing activities carried out. You could choose to sample some aspects which have been IQAd and some which have not

- minutes of meetings and standardisation records to check all aspects are being covered over time.

Always ensure you have read a copy of the last EQA report and any action and improvement points, as this will form the basis of your activity. Improvement points are for guidance, whereas action points are monitored and enforced, and can affect a centre's risk rating (see Chapter 3) if they are not met by an agreed date.

If the centre is new, there will be a report based on the approval process which may still contain some action points which must be followed up. Ideally, if you keep in touch with your centres between monitoring activities, you will have been kept up to date with their progress regarding any action points.

Based on the information you obtain, any action points and risks associated with the centre, you will be able to plan what you need to sample and when. You will also need to decide what other documents you will need to see, for example, policies, procedures, records of appeals and complaints. When planning what to sample, you should take into account learners' work that has been assessed and which has not been internally quality assured, as well as work that has been assessed and has been internally quality assured. This should be from current and previous learners to give you a good idea of how the centre operates.

You might need to adapt your planning and sampling approaches to meet the needs of your centre or if you find a problem. This might include providing help and support, challenging a decision an assessor has made or carrying out an additional random sample if you have concerns about something. Whatever happens, you should never compromise quality or take anything personally; you should always remain objective with your decisions and act professionally.

Preparing for a centre visit

You will be able to use the information you have received from the centre and the AO to create a visit and sample plan document (see Table 4.2). Most awarding organisations will supply templates for you to use and you will need to know how to access and complete them.

There are different types of centre visits that you might be required to carry out, for example, to approve a new centre to offer qualifications or to sample from an existing centre. These will be explained as you progress through the chapter.

Your plan should also include a list of your activities and approximate timings, for example, when you wish to see certain staff members. However, some AOs don't expect you to add times, as some activities might take longer than others. Examples of activities you could carry out are covered in the next section of this chapter.

Make sure you allow time for conversations, questions from staff, additional random sampling (in case you identify a problem), report writing and feedback. If you need anything specific such as an internet connection, make sure you ask well in advance.

Once you have completed your visit and sample plan, you will need to send it to the centre (and you might need to send a copy to the AO) within the required timescale. Nearer to the activity, you will need to:

- confirm the date and time of the activity with the centre

- check the centre has received your plan

- ensure you have personal identification such as a badge or an AO identification card

- check public transport or route plans and parking arrangements

- prepare all the necessary documents, checklists and reports you will use (these might be accessible manually or electronically).

A similar form to Table 4.2 could be used for remote monitoring activities which do not include a visit to a centre.

Table 4.2 Example visit and sample plan

Centre name:	XYZ Company	Centre number:		0123456
Centre contact name:	Jean Haverham (IQA)	Location/s to be visited:		Manor Building Main Road MillerVille
External quality assurer's name:	Paulo Hancock	Contact details:		01234 234567
Date of visit:	5 Sept	Time of arrival and duration of visit:		9.00 am Approx 6–7 hours
Qualification/s to be sampled:	Level 2 Customer Service Certificate			
Staff to meet:	Jean Haverham Phil Scholey Terri Mandalay		Internal quality assurer and centre co-ordinator Trainer/Assessor Trainer/Assessor	
Learners to meet:	Marie Maine			
Activities to be carried out with approximate timings:	9.00	Arrival, meet team and discuss progress since approval, answer any questions. Check internal quality assurance systems and procedures		
	9.30	Observation of Phil delivering a training session in the centre to a group (I will just observe the first half hour, then talk to a few learners)		
	10.00	Travel to workplace then carry out an observation of Jean observing Terri with her learner Marie Maine		
	11.00	Talk to Terri and Marie, then return to centre		
	12.00	Interview with Jean and Phil		
	12.30	Commence sampling learners' work and associated assessment and IQA records, whilst having a working lunch. Sample minutes of meetings, standardisation records, staff CPD records and updated policies and procedures		
	3.00	Complete report and provide feedback to Jean, answer any questions and discuss any action and improvement points		
	3.30	Depart		
Learners' work to be sampled, along with associated assessment and IQA records:	Pierre Smithson Jeremy Globe Jo Trotter Melanie Green Frank Glass Jeremiah Marsh		Unit 1 Units 1 and 2 Units 1 and 3 All units completed to date All units completed to date Units 1, 2 and 3	
Additional comments:	Please check the last report and ensure there is evidence that all action points have been met I may carry out an additional random sample on the day			
Date form sent to the centre:	5 August			

For the purposes of future proofing this textbook, a year has not been added to dates in the tables, reports and checklists. You should always add the year as well as the day and month to any records you complete.

Extension activity

Obtain the visit and sample planning documentation that you will use for your next centre visit. If you are about to plan an EQA monitoring activity, obtain relevant information from your centre and complete the documents. Ensure you can sample a good cross section of work and records from learners, assessors, IQAs, qualifications and locations.

Find out the final registration and certification dates for the qualifications you will monitor. If they are within the year, obtain a copy of the replacement qualification specification and make sure your centre is aware of any changes.

External quality assurance activities

External quality assurance activities are not just about meeting with staff and sampling learners' work. Read (2012: 104) states: *The EQA checks that the IQA process is fit for purpose and samples evidence to show that it is being implemented appropriately. For example, you would check that observed assessments are carried out in line with the IQA policy, and that any problems identified are followed up and resolved.*

The EQA activities should also include:

- checking policies and procedures are up to date
- observing trainers', assessors' and internal quality assurers' practice
- sampling documents such as minutes of meetings and standardisation records
- talking to learners and others (e.g. workplace witnesses).

Whichever activities are chosen, they should be fit for purpose and you should ensure that something from each internal quality assurer and each assessor is sampled over time. You will need to keep a note of whom in you have sampled from and when. It's about quality, not quantity. The size of the sample will be based on the centre's risk rating and your planned strategy.

You should always remain professional and not let any personal issues affect the sampling process or the activities you carry out. For example, you should not expect a centre to do something that is not a requirement, e.g. create and use a particular document, or carry out an observation of a learner a certain number of times if it's not a requirement. When asking a centre to do something, it must always be based on a written requirement or regulation, and not just be your opinion. However, you can offer suggestions if you have ideas to help your centre improve, perhaps based on your own experiences of working in a centre. Centres are able to make their own decisions regarding how they facilitate the qualifications. As an EQA, you cannot impose your own personal preferences upon them as long as they are meeting the relevant requirements, regulations and standards. If you do have any suggestions, they will be classed as *improvement* or *development* points which are just recommendations and not *action* points. Improvement and development points are for guidance only and are not enforceable. Action points are based on requirements which must be monitored and enforced. Action points can affect a centre's risk rating if they are not met by an agreed date, whereas improvement and development points will not.

Your awarding organisation will inform you how regularly you will need to visit a centre, usually based upon their risk rating, i.e. low, medium or high (see Chapter 3). A low-risk centre might be operating well, and you could therefore carry out a remote monitoring activity instead of a visit. This involves the centre posting documents to you (or making them accessible electronically) rather than you visiting them.

Never arrange any activity without obtaining the approval of the awarding organisation first, as, depending upon your contract, you might not be paid for any work which has not been authorised. You might also not get paid for any activities you carry out until you submit a report to the AO. A separate claim for any expenses such as travel costs might also need to be submitted, along with receipts.

Examples of EQA activities

The different activities you might be required to carry out include:

- an approval visit for an organisation to become a centre

- an approval visit for an organisation to offer an additional qualification

- an advice or support visit to give guidance and/or training

- a systems' visit to check quality assurance systems, policies and procedures

- a monitoring visit (for particular qualifications) to meet staff, sample learners' work, observe practice and sample other relevant documents and records

- a remote monitoring activity to sample learners' work, assessment and internal quality assurance documents.

The types and titles of these activities will differ between awarding organisations and are explained after the following activity. Once it has been agreed with the awarding organisation for you to carry out an activity, you can confirm your requirements to the centre or they might confirm them on your behalf. Some activities are free of charge for centres, whereas others are chargeable by the AO.

Activity

Find out from the AO what the different types of activities are that you will be required to carry out and what is involved for each. Ascertain if a centre will be charged for any, and if so, how much this will be. You will then be able to advise your centres if asked.

Approval visit for an organisation to become a centre

All centres have to start somewhere and a new centre will need an approval visit to ensure they have systems and procedures in place to operate effectively.

If a centre is currently accredited via another awarding organisation and has a centre number, the process should be quite straightforward as they will have recognised systems and procedures in place. Your awarding organisation might ask you to view at least two recent EQA reports from the other AO, to check that they are being compliant and do not have any sanctions or action points outstanding. You might like to find out why the centre wants to change to a different AO, in case something has happened, for example, an argument with their previous EQA.

If the centre is not accredited with another AO, you will need to carry out a thorough check of all their systems, procedures and resources to ensure they meet the relevant requirements, regulations and standards. Rather than looking at everything first when you arrive, and becoming overwhelmed, you could ask your centre contact to set the scene for you. They could describe how their centre operates and explain to you how their policies and procedures work. This will help you get an overall picture before looking at the documentation. Your AO will give you guidance as to what you need to see and do.

Aspects to carry out for a centre approval include:

- meeting relevant staff, such as managers, administrators, internal quality assurers, trainers and assessors;

- looking at policies and procedures to ensure they meet all requirements, such as appeals, assessment, complaints, evaluation, equality and diversity, health and safety, initial assessment, quality assurance, safeguarding;

- looking at the assessment and internal quality assurance systems and documents to ensure they meet all requirements;

- seeing the record-keeping system, whether manual or electronic, and ensuring data protection and confidentiality requirements will be met;

- ascertaining details of other locations, for example, different sites or workplaces where training and assessment will take place.

During a centre approval visit, you can take the opportunity to discuss other aspects. These might include use of the awarding organisation's logo, malpractice, the AO's guidance for registration and certification of learners, accessing updates via their website and anything else that is relevant at the time.

Based on the previous bullet points, and anything else you have reviewed, you will need to make a decision as to whether the centre can be approved or not. You will need to complete a centre approval report which will be supplied by your awarding organisation. The report might include some action points, such as to update and e-mail you a revised appeals procedure, which the centre will need to address by a set date. This report might be given to the centre or go to them via the awarding organisation. You will need to find out what the process is.

If the centre does not meet all the relevant requirements, regulations and standards, and there are several action points, an advisory visit might be necessary to give further advice and support before centre approval is granted.

Approval visit for an organisation to offer additional qualifications

An existing centre might need an approval visit to add another qualification to their portfolio. You could carry out a qualification approval at the same time as a centre approval, providing everything you need to see is in place. Your awarding organisation will give you guidance as to what you need to see and do.

Aspects to check for qualification approval include:

*   checking relevant staff's CVs, continuing professional development (CPD) records and original qualification certificates to ensure they meet the requirements to deliver, assess and internally quality assure the qualification

*   looking at resources such as training rooms, equipment, library and computer facilities to ensure they meet the requirements of the qualification

*   reviewing particular documents that relate to the qualification, for example, assessment and internal quality assurance policies, strategies, templates, training materials, recruitment and marketing materials, schemes of work and session plans if relevant

*   reviewing the initial assessment, induction, progress review and evaluation procedures and documents.

Based on the bullet points, and anything else you have reviewed, you will need to make a decision as to whether the centre can be approved or not to offer the additional qualifications. You will need to complete a qualification approval report which will be supplied by your awarding organisation. The report might include some action points, such as to update and e-mail you the revised initial assessment procedure, which the centre will need to address by a set date. This report might be given to the centre, or go to them via the awarding organisation. You will need to find out what the process is.

If the centre does not meet all the relevant requirements, regulations and standards, and there are several action points, an advisory visit based on the particular qualification requirements might be necessary to give advice and support before qualification approval is granted.

Activity

If you are asked to carry out a qualification approval visit to a centre, how can you ensure the staff and resources meet the relevant requirements, regulations and standards? What would you be looking for regarding your particular subject area?

Advice or support visit to give guidance and/or training

These types of visit are often referred to as an *advisory visit* and can take place at any time. They can be requested by the centre, or decided by the awarding organisation if they feel the centre needs support. A visit could take place prior to a centre or a qualification approval activity or at any time the centre is operating, for example, if a centre feels they need some particular advice or training, perhaps when qualification criteria change. It might also be when a centre has been placed on a serious sanction and needs some staff development. Your awarding organisation will give you guidance as to what you need to see and do, and you will need to complete a report as to the activities you carried out. This report might be given to the centre, or go to them via the awarding organisation. You will need to find out what the process is.

Systems' visit to check quality assurance systems, policies and procedures

A systems' visit might take place separately from an EQA monitoring visit. This type of visit is to check the management and quality systems in place at the centre, rather than to sample learners' work. This might include checking how relevant policies and procedures are put into practice, how learners are recruited, inducted and supported, the resources which are available and how records are stored and accessed.

A person other than the subject specialist EQA might carry out this visit, or it could be by the usual EQA. This type of visit might take half a day and be followed up on another occasion by an EQA monitoring visit. The latter will then focus on sampling specific qualifications rather than systems and procedures. Your awarding organisation will give you guidance as to what you need to see and do, and you will need to complete a report as to the activities you carried out. This report might be given to the centre, or go to them via the awarding organisation. You will need to find out what the process is.

Monitoring visit for particular qualifications

This type of visit is to meet staff, sample learners' work, observe practice and sample other relevant documents and records. It can be very time consuming to obtain the relevant information in advance to plan for your visit. A visit can also take several hours, plus travelling time, depending on what you need to do and where you need to go. Systems and procedures might be checked during this visit if a separate systems' visit has not already taken place.

It might be that you have not met the centre staff previously and therefore don't know much about how they operate. Rather than looking at everything first when you arrive, and becoming overwhelmed, you could ask your centre contact to set the scene for you. They could describe how their centre operates and explain to you how their policies and procedures work. This will help you get an overall picture before commencing your monitoring activities.

Never feel obliged to sample what the centre staff ask you to, as you will only ever see what they want you to see and you could easily miss something that isn't meeting the requirements. It is useful to carry out an additional *random sample* on the day. This could be to check if a trend is occurring, for example, if all assessors are interpreting the requirements incorrectly.

Example

Xavier, the external quality assurer, had chosen to sample Unit 4 from two different learners who had been assessed by the same assessor. In both samples, the assessor had made an incorrect decision. Xavier decided to carry out a further random sample of the same unit, but from two other learners, who had been assessed by a different assessor. He noticed the second assessor had also made the same mistake. This identified to Xavier that both assessors needed guidance. This was a serious matter as the internal quality assurer had not noticed this when carrying out her sample. Fortunately, the learners were still working towards their qualification and could be re-assessed. However, they had been disadvantaged due to the assessors' mistake. Xavier made a note to add an action point to his report regarding this.

During the visit

You will have sent a plan to the centre in advance, which outlines what you would like to see and do. However, the timings of these and the actual activities might change due to your findings during the day. You need to remember that you are in control of the visit and the activities you wish to carry out. You should not be persuaded by any centre staff to do anything other than what you feel is relevant and appropriate.

Activities you could carry out during a visit include:

- observing teaching and training sessions

- observing assessment practice and the feedback given to learners

- watching an internal quality assurer observing an assessor, then giving feedback

- observing a standardisation activity taking place

- talking to staff, learners and other people such as workplace witnesses

- sampling learner work which has been assessed formatively and summatively, and which has and has not been internally quality assured

- sampling minutes of meetings, standardisation records, appeals and complaints, initial and diagnostic assessment results, interview and induction records, tutorial review records, analysis of programme evaluations and questionnaires

- reviewing statistics (such as number of starters and completers, the amount of time taken between learners starting and completing the full qualification, the time taken between learners commencing and being registered with the AO, success and progression rates for the programme)

- monitoring policies and procedures and when they were reviewed/updated

- discussing Ofsted or other report findings if applicable.

You will have notified your centre on your plan of who you wish to see and when, so that they can be available. If a learner or their work is not available on the day, the centre should have informed you beforehand, so that you can choose someone else, rather than the centre choosing for you. If the centre chooses, this could be because they don't want to bring your attention to an issue. What you could do in this situation is ask to see the assessor's records for the learner. This will prove the learner does exist, and should show the progress the learner has made so far.

Your awarding organisation will give you guidance as to what you need to see and do. However, it will be down to you to decide how much you will sample and who you see. This will be based on your knowledge of the centre, any outstanding action points and any risks you feel could disadvantage the learners. You will need to complete a report as to the activities you carried out. This report might be given to the centre or go to them via the awarding organisation. You will need to find out what the process is.

The following checklist might help you during a centre visit.

External quality assurance checklist

- ☐ *Arrive on time, have your identification available and/or wear a name badge, sign in as necessary at the centre and follow their safety and security procedures.*

- ☐ *Act professionally at all times and remain objective with your judgements.*

- ☐ *Follow all AO and regularity requirements.*

- ☐ *Be helpful and supportive.*

- ☐ *Keep records of everything you do i.e. observations, discussions, sampling (either on the AO's documentation or your own).*

- ☐ *Follow your plan, however this might change as you progress through the visit, e.g. due to staff commitments or depending upon what you find.*

- ☐ *Carry out additional random samples if you identify any risk, e.g. incorrect assessor decisions or plagiarism.*

- ☐ *Talk to learners; you might get a different perspective to that given to you by assessors and internal quality assurers.*

- ☐ *View training rooms, learning resource and library facilities, other resources and equipment as necessary.*

- ☐ *Contact the awarding organisation if you identify a serious risk, e.g. malpractice (do not alert the centre at this point but seek AO guidance).*

- ☐ *Remain objective with your decisions and back up everything by the book, i.e. don't ask a centre to do something which is not a written requirement.*

- ☐ *Complete the awarding organisation report accurately and legibly.*

- ☐ *Identify an appropriate risk rating, usually low, medium or high.*

cause for concern, as these issues should have been notified to her beforehand. Annabelle then telephoned the awarding organisation, who told her not to go ahead with the sample but to wait until the centre could send her what she had originally requested, plus a further sample. In the meantime, the centre was placed on a sanction so that they could not claim any certificates.

Not everything will always go smoothly when carrying out a remote monitoring activity, as you won't have direct access to staff, learners and documents. However, if you have ongoing communication with your centre, most issues can be resolved before they become serious. It's about being proactive not reactive.

Your awarding organisation will give you guidance as to what you need to see and do. However, it will be down to you to decide how much you will sample and who you see. This will be based on your knowledge of the centre, any outstanding action points and any risks you feel could occur to disadvantage the learners. You will need to complete a report as to the activities you carried out. This report might be given to the centre or go to them via the awarding organisation. You will need to find out what the process is.

Sampling learners' work

When sampling learners' work, whether during a visit or a remote monitoring activity, you are not re-assessing or re-marking it. You are checking if the assessor has adequately planned and documented the assessment activities, made a correct decision and given constructive and developmental feedback. Otherwise, the learners might be disadvantaged due to no fault of their own.

When sampling work from different assessors, if you are sampling the same aspects you can see how consistent the different assessors are. You can then note any inconsistencies and feed this back to the centre staff. For example, if one assessor is giving more support to learners, or expecting them to produce far more work than others, then this is clearly unfair. Some assessors might produce detailed assessment plans and others might be minimal; the same might apply with the amount of feedback given. You need to ensure the staff are consistent by standardising their approach. If not, then this would become an action point for the centre to address.

Sampling learners' work is also a good opportunity to check for other things, for example, if plagiarism or copying is taking place. You should also check whether the learners have had the opportunity to be assessed in another language which is acceptable, for example, bilingually (e.g. English and/or Welsh or Gaelic). Some AOs allow the use of an interpreter, but others don't; you would therefore need to find out what is acceptable. Other aspects to check include the use of holistic planning and assessment. There's no need for an assessor to observe all the different units separately, if one or two well-planned holistic observations will do. Another area to check is the recognition of prior learning (RPL). If a learner already holds an approved or accepted unit from another qualification, they should not have to repeat it.

Because you are only sampling aspects of the assessment and internal quality assurance process, there will be some areas that get missed. This is a risk as you can't sample every thing from everyone. You need to build up your confidence in the centre's staff to know they are performing satisfactorily. If you find a problem when sampling, or have any concerns, you will need to increase your sample size. You can always carry out an additional random sample and ask to see something which isn't on your original plan. However, your decision can only be based upon what you have sampled and seen, not on any feelings you might have which are not substantiated.

Make sure you keep notes of what was sampled and update your report as you progress. If you are completing an electronic report, make sure you save it regularly and make a backup copy.

Documents for sampling activities

The awarding organisation might supply you with documents, checklists and reports to use for your sampling activities. This way, you have a formal record of what was carried out and when. If they don't, you could design your own based on those here, to ensure you are keeping full and accurate auditable records.

Examples include:

- observation checklist for internal quality assurer practice
- observation checklist for training delivery (only required if this is part of the process to be monitored for a particular qualification)
- observation checklist for assessment practice
- interview checklist with an internal quality assurer
- interview checklist with a trainer/assessor
- interview checklist with a learner
- interview checklist with a witness.

The following tables give examples of the checklists in the above list. They have been completed as though an external quality assurer has used them. Records can often be completed electronically and might not need a signature. You should check with your awarding organisation what checklists they provide, if signatures are required and how they expect you to complete, distribute, save and/or file them. You will find they are similar to ones that IQAs could use. All documents should include the year with the dates.

Extension activity

Find out which documents, templates and reports are available for you to use for your EQA role. Are they in hard-copy format or can you access and use them electronically? What other documents will you need to use which might not be supplied by the AO, for example, various checklists like those in the following examples? What are the awarding organisation's timescales for completing and submitting the various documents and reports?

Table 4.3 Example observation checklist for internal quality assurer practice

Observation checklist for internal quality assurer practice		
IQA: Jean Haverham	EQA: Paulo Hancock	
Assessor: Terri Mandalay (trainee assessor)	Date: 05.09	
Checklist	**Yes No N/A**	**Comments/responses/action required**
Did the IQA put the assessor at ease and explain what they were going to observe?	Y	Jean explained she was giving feedback to Terri regarding her assessment decision for unit 4. This unit had been checked and countersigned by Phil as Terri is working towards the assessor qualification.
Did the IQA give feedback to the assessor in a constructive and developmental way?	Y	Jean was very detailed with her feedback and gave advice as to how the updated assessment forms should be completed according to company requirements.
Were appropriate IQA records used?	Y	Jean referred to her IQA sampling plan and tracking sheet, and the IQA report for unit 4. She also had a copy of the Customer Service qualification specification.
Were questions appropriate and asked in an encouraging manner?	Y	Jean asked open questions to ensure Terri could give appropriate responses. Jean's manner was encouraging and she came across as very approachable.
Was the assessor encouraged to ask the IQA questions and clarify points?	Y	Terri felt able to ask questions, and did interrupt Jean on occasion with very pertinent questions.
Did the IQA give adequate support to the assessor?	Y	Good support was given, particularly regarding record keeping and the company requirements.
Was the IQA's decision correct?	Y	Jean's decision supported Terri's judgement.
Did the IQA agree any action points or development points with the assessor?	N	The opportunity should have been taken to discuss Terri's progress towards the assessment qualification.
Were all IQA records completed correctly?	Y	The IQA sample plan, tracking sheet and reports were all completed.
Did the IQA perform fairly and satisfactorily?	Y	Jean performed well considering I was observing the discussion.
Does the IQA have any questions?	Y	Q – Do I need to retake my IV qualification now that it's known as IQA? A – No – as long as you are demonstrating the requirements of the current IQA standards.

Feedback to IQA:
This was a very good feedback session. You put Terri at ease and guided her through the requirements of assessment of the unit and your company's systems. Don't forget to ask how she is getting on with her assessor qualification.

Table 4.4 Example observation checklist for training delivery

Observation checklist for training delivery		
Trainer: Phil Scholey	EQA: Paulo Hancock	
Units/aspects being observed: Unit 3	Date: 05.09	
Number in group: 10		
Checklist	**Yes No N/A**	**Comments/responses/action required**
Were the aims/objectives/learning outcomes clearly introduced?	Y	The aim and clear objectives were stated and kept on display throughout the session. Phil explained how they related to the learning outcomes of the unit.
Do learners have action plans/individual learning plans?	Y	All learners have individual learning plans to denote units and target dates for assessment.
Is there a scheme of work and session plan?	Y	Both documents were available and relevant to the session being taught.
Are the resources and environment safe and suitable?	Y	Classroom is suitable. Resources are adequate and plentiful for the group size.
Does the session flow logically? Are the learners actively engaged?	Y	Topics flowed logically. Learners were involved in group activities, paired and individual work.
Are open questions used to check knowledge?	N	Closed questions were used too often – try to use open questions more.
Are all learners able to ask questions and contribute to the session?	Y	Learners were encouraged to ask questions throughout the session.
Was learning taking place?	Y	Several different activities were used to observe skills. The knowledge and understanding of each learner was checked by the use of individual questions, therefore learning was taking place.
Was clear feedback given regarding progress to each learner?	Y	Group feedback was given after each activity. Individual feedback was given throughout the session.
Was the session formally summarised?	N	No summary took place linking to the original objectives of the session.
Was the next session explained? (If applicable)	Y	Topic to be covered next week was clearly explained.
Did individual learners know what they had achieved during the session and how it relates to their qualification/ programme of learning?	Y	Feedback linked achievements to the qualification's unit for each learner.
Are all relevant records up to date?	Y	I checked the course folder – all relevant records were seen, including the group profile.
Do any learners require additional support?	Y	No learners had disclosed any particular needs – trainer could tactfully ask them
Did the trainer perform fairly and satisfactorily?	Y	The trainer performed fairly and satisfactorily. He covered all the objectives.

Feedback to trainer:
This was a well-planned and delivered session, however don't forget to involve your learners by asking open questions (e.g. ones that begin with: who, what, when, where, why and how). The group was small, therefore you were able to give individual attention and feedback. You do need to summarise your session at the end and link to the original objectives.

Table 4.5 Example observation checklist for assessment practice

Observation checklist for assessment practice

Assessor: Terri Mandalay (trainee assessor)　　　　EQA: Paulo Hancock

Units/aspects being assessed: Unit 5　　　　Date: 05.09

Group size or individual assessment: Individual

Checklist	Yes No N/A	Comments/responses/action required
Was the learner put at ease and aware of what would be assessed?	Y	You prepared the learner well by explaining what you would observe and how you would do it; this helped relax them.
Was an appropriate assessment plan in place?	Y	You had a detailed assessment plan in place which had been agreed in advance.
Were the resources and environment healthy, safe and suitable for the activities being assessed?	Y	All resources were appropriate and suitable. There were no issues with health and safety.
Were questions appropriate and asked in an encouraging manner to all learners?	Y	You asked open questions to confirm knowledge and understanding.
Were current and previous skills and knowledge used to make a decision?	Y	You took into account the fact your learner had already achieved an aspect of the unit prior to this assessment.
Was constructive and developmental feedback given and documented?	Y	Feedback was very positive and constructive, you asked your learner how they felt they had done – this prompted a few questions which led to a discussion regarding further development.
Was the assessor's decision correct?	Y	Your judgement was correct and accurate.
Did the learner's evidence meet VACSR requirements? (valid, authentic, current, sufficient and reliable)	Y	You checked VACSR for all the aspects you assessed and there were no issues.
Were all assessment records completed correctly?	N	Complete the observation report as soon as you can after the observation and give a copy to your learner.
Did the assessor perform fairly and satisfactorily?	Y	Yes, apart from filling in the observation report due to time limits.
Does the assessor have any questions?	Y	Q – How long do my decisions need to be countersigned? A – Until you are qualified as an assessor.
What assessment activities were used and why?	Observation – to confirm performance. Questions – to check knowledge and understanding.	

Feedback to assessor:
I liked the way you were unobtrusive during the observation, yet asked open questions to confirm knowledge as the learner carried out their job role. I appreciate you did not want to fully complete the observation report due to time constraints. However, you did confirm verbally how your learner had done. Once the observation report is completed, please give a copy to your learner.

Table 4.6 Example interview checklist with an internal quality assurer

Interview checklist with an internal quality assurer	
IQA: Jean Haverham	EQA: Paulo Hancock
Qualification: Customer Service	Date: 05.09
Checklist	**Comments/responses/action required**
How long have you been internally quality assuring this qualification?	Nine years at this organisation and three years at my previous organisation – I have seen three changes in standards.
What qualifications do you hold? e.g. IQA and subject specific qualifications (EQA to see certificates/CVs)	Level 4 Customer Service Level 4 Learning and Development D34 Internal Verifier Award and I have attended an IQA update session.
How do you maintain your continuing professional development?	I take part in the company's in-house events for assessors and IQAs, as well as standardisation activities. I recently attended an external CPD session regarding new technology.
How many assessors are you responsible for and how do you allocate learners to them?	Two – Phil is experienced and qualified. Terri has been working in customer service for many years but is new to assessing. Phil is currently countersigning Terri's decisions. Terri has fewer learners than Phil until she becomes qualified as an assessor. They are allocated according to location.
How do you induct new assessors and what documents do you give them?	I have a meeting with them and use an induction checklist to ensure I cover all the necessary points. Each assessor is given a copy of the company handbook.
What is your sampling strategy for this qualification?	• Observe each assessor every six months. • Talk to a sample of learners and witnesses. • Sample all units from each assessor across a mix of learners over a period of time (new assessors will have a higher sample rate). • Chair a team meeting and standardisation activity every other month. • Maintain full records of all IQA activities. • Implement external quality assurance action points.
Which assessors have you observed since my last visit and why? (EQA to see records)	I observed Phil two months ago, and Terri one month ago. I have a plan which shows the dates I plan to observe them.
Have you formally spoken to learners about the qualification process? (EQA to see records)	Yes, I always speak to at least two learners from each assessor and I use a learner interview checklist to document this.
Have there been any appeals or disputes against your assessors or yourself?	No – we did have an appeal against an assessor for the old standards, but he has now left the organisation.
Have there been any changes to staff/ resources since my last visit?	Since your approval visit there have been no changes.
How often do you hold team meetings and standardisation activities? (EQA to see records)	Every other month, as in my IQA strategy. However, if I find a problem I might hold another meeting. Records are kept of all units standardised and minutes are kept of meetings.
Do you have any questions?	Q – When are you due to visit again? A – This will depend upon learner progress, however I might carry out a remote sample rather than a visit. Please keep in touch to update me on progress.

Table 4.7 Example interview checklist with a trainer or assessor

Interview checklist with a trainer or assessor	
Assessor: Phil Scholey	EQA: Paulo Hancock
Qualification: Customer Service	Date: 05.09
Checklist	**Comments/responses/action required**
How long have you been training/assessing this qualification?	Eight years – I assessed the previous version before it recently changed.
What qualifications do you hold? (EQA to see certificates/CVs)	Customer Service, Management and Retail Diplomas at level 4. A1 Assessor Award with an update to ensure I am working at the current standards.
How do you maintain your continuing professional development?	The company organises training events on various topics. The last one was about e-assessment. We also have regular standardisation activities.
How many learners do you have? Are they based at different sites?	Ten on each level. They all come into the organisation for knowledge training. They each have a different place of work which is where I assess their performance.
How do you induct your learners and what documents do you give them?	There is an induction checklist to ensure all points are covered. Each learner is given an electronic copy of the learner handbook and the qualification specification.
What documentation do you use for the assessment process and why?	Assessment plans, observation checklists and questioning records – to ensure all aspects of the units are planned and assessed according to the qualification requirements.
How do you take into account prior learning and experience?	By discussion with the learner and viewing evidence provided by them. I try and assess holistically when I can e.g. where aspects of units overlap.
How often do you review the progress of each learner?	Every six weeks in the workplace.
Have there been any appeals or disputes against your decisions?	Not that I am aware of.
How often do you attend team meetings and standardisation activities? (EQA to see records)	We have a team meeting and standardisation event about every eight weeks. Records are electronic.
How is your practice monitored?	I'm observed assessing and giving feedback every other month. The IQA talks to my learners and samples my records.
Do you have any questions?	Q – When will the qualification be updated? A – I'm not sure – I will find out and get back to you.

Table 4.8 Example interview checklist with a learner

Interview checklist with a learner	
Assessor: Terri Mandalay (trainee assessor) Learner: Marie Maine	EQA name: Paulo Hancock Date: 05.09
Checklist	**Comments/responses/action required**
What have you achieved so far?	Unit 105
How do you discuss and agree your assessment plan? Is anyone else involved?	I discuss with my assessor what she will look at, which is based on what I'm covering in the workplace. She often talks to my supervisor who is classed as a witness.
Did you have a copy of, and understand, what you are being assessed towards?	I have a copy of the qualification specification and assessment for unit 5 was fully explained to me.
Did you have an initial assessment and/or were your previous skills and knowledge taken into account?	I had told Terri what had been achieved previously – she took this into account and checked my evidence.
Were you asked questions to check your knowledge and understanding?	Yes, Terri links questions regarding my knowledge to my practice at work.
Did you receive helpful feedback?	Yes, after the assessment. Note: this was verbal but needs to be formally documented – assessor to follow up.
How is your progress regularly reviewed?	Once every 6–8 weeks.
What opportunities are there to progress further?	Learner wasn't sure.
If you disagreed with your assessor, would you know what to do?	No. Note: assessor to explain the appeals procedure.
Do you have any learning needs or require further support?	The learner is happy with the support given.
Do you have any questions?	Q – When do I get my certificate? A – Three weeks after successful achievement, providing the centre claims it on time.
Feedback to assessor after discussion with the learner: Your learner was very pleased with the way the assessment was conducted. However, she is unsure of the appeals procedure. Please direct her to the learner handbook which contains a copy. You should also chat to her about progression opportunities for when she completes her qualification.	

Table 4.9 Example interview checklist with a witness

Interview checklist with a witness

Witness: Poppy Lane

Learner: Cherie Lowrie

EQA name: Paulo Hancock

Date: 05.09

Checklist	Comments/responses/action required
Are you aware of the requirements of the qualification?	Yes, I have a copy of the qualification specification.
What training has the centre given you regarding being a witness?	I had one day at the centre with witnesses from other organisations. We all discussed the requirements of the units and how we would complete the witness testimony documents.
How do you confirm your learner has the required skills and knowledge to meet the qualification criteria?	I am Cherie's supervisor so I work closely with her. I can observe her performing skills and I ask her questions to confirm her knowledge and understanding.
How do you write your witness testimonies?	I am quite detailed when writing and I cross-reference what Cherie has done with the assessment criteria from the qualification. I never complete a witness testimony unless Cherie has achieved all the aspects required. I save them electronically and email to Cherie.
How do the centre staff communicate with you?	I get a phone call once a week from the assessor at the centre. It gives us a chance to chat about how Cherie is progressing and link theory and practice.
How does the off-the-job training tie in with the workplace?	The assessor and I talk about Cherie's day release training and how it fits in with what she's currently working on. I also inform the assessor what tasks Cherie is learning here, so that they can cover a bit more of the theory side of it at the centre.
Do you feel you have the time and the expertise to devote to being a witness?	I am lucky in that my organisation is keen to take on apprentices and give people like me the time to support them.
Does your learner have the opportunity to progress further?	Yes, she will be kept on full time after her apprenticeship and will have the opportunity to partake in further training and development.
Is there anything you would like to comment on that we have not discussed?	I'd like to say how well Cherie is progressing.
Do you have any questions?	Q – Is there is a qualification I can take to help support apprentices in the workplace? A – I will find out what is available and ask the centre to liaise with you.

Feedback to witness:
It was good to meet you today and I'm really pleased Cherie is progressing well, and that the centre is supporting you. Thank you very much for your time.

Making decisions

You will need to make several decisions regarding what you have sampled and seen during your monitoring activities. These will be based on whether the centre staff have performed satisfactorily and followed the requirements of the qualification, the awarding organisation and any relevant regulations. As the learners are registered with your awarding organisation, you have a duty to ensure they are being treated fairly and not being disadvantaged in any way. You also have to uphold the credibility of the qualification and reputation of the awarding organisation. Your decisions must always be based on facts.

Making a decision: observing practice

A good way of ensuring assessment and IQA practice are adequate is to see this in action. Not only will this give you the opportunity to see how decisions are made, but you will also be able to talk to the staff and learners afterwards. Using a checklist like those in the previous section of this chapter will help you to focus on various aspects and enable you to document what you have observed and discussed.

If what you see meets the requirements of the qualification, you can document this in the awarding organisation report. If not, you will need to explain to the centre staff what needs correcting and agree an action point which must be added to the report with a target date.

Making a decision: sampling assessed work

You need to sample according to your plan (Table 4.2), however you can sample more if you need to, for example, if you find something wrong. You need to confirm what you have sampled was robust, safe, valid, fair, reliable and ethical.

- Robust – the assessment activities used are vigorous and will endure the test of time.

- Safe – there is little chance of plagiarism by learners, the work can be confirmed as authentic, confidentiality is taken into account, learning and assessment is not compromised, nor is the learner's experience or potential to achieve (safe in this context does not relate to health and safety but to the assessment methods used).

- Valid – the assessment activities used are based on the requirements of the qualification.

- Fair – the activities used are appropriate to all learners at the required level, taking into account any particular learner needs. All learners should have an equal chance of an accurate assessment decision.

- Reliable – the assessment activities used would lead to a similar outcome with similar learners.

- Ethical – the assessment takes into account confidentiality, integrity, safety, security and learner welfare.

Activity

What do you feel might affect the decisions you make, either when visiting a centre or carrying out a remote monitoring activity? Do you think you could be influenced by anyone or anything for a particular reason? If so, how would you deal with the situation?

When sampling, you might find the following checklist (in alphabetical order) useful to help you make decisions and monitor risk.

Sampling checklist

☐ *Access – is the centre preventing you from accessing certain records, people or locations? If they cannot give a valid reason then this is a cause for concern.*

☐ *Action from the previous report – has this been completed? If not, why not? If the centre has not communicated their reasons to you it might place them on a higher risk rating.*

☐ *Appeals and/or complaints received – why is this happening? Is there a pattern? Is the centre's policy and procedure adequate?*

☐ *Assessment methods and types of evidence provided by learners – does it meet the requirements? Are any assessors over or under assessing? Do learners have adequate assessment plans/action plans? Do they receive developmental feedback? Is initial, formative and summative assessment correctly carried out? Have assessors used appropriate or alternative methods, for example, asking oral questions rather than issuing written questions for a learner who has dyslexia? Is holistic assessment carried out where possible? Is recognition of prior learning (RPL) used and documented where applicable? If written questions are developed by the centre, are sample answers used to standardise expected responses? Is equality and diversity taken into account? Are special assessment requirements taken into account where necessary?*

☐ *Assessment strategy and requirements for assessors to be qualified – has everyone interpreted these in the same way? Are all assessors experienced and/or qualified in their subject area? Are they countersigned if they are working towards an assessor or IQA qualification (if required)? Are staff changes notified to you and the awarding organisation?*

☐ *Authenticity of learners' work – do you suspect plagiarism or cheating? Have all learners signed a document to state the work is their own? If the work is created and stored electronically, can you be certain it is authentic and can it all be attributed to the learner?*

☐ *Awarding organisation guidance and regulations – is the centre aware of and following these? Is the centre receiving regular updates and information (for* example via the awarding organisation's website) and discussing these with relevant staff?

☐ Centre – has it been established a while or is it new? Have they offered this qualification for a long time, i.e. how experienced are they, or are they becoming complacent?

☐ Certification – if the centre has direct claim status, are they claiming the certificates according to the awarding organisation's requirements and in a set time after completion? Are certificates being claimed before learners have been assessed and internally quality assured? Are certificates issued to learners in a timely manner?

☐ Communication – how do the staff communicate with you (between visits and during a visit)? Are they defensive when you ask questions or open to listening to you?

☐ Consistency – are internal quality assurers being fair to assessors, and assessors fair to learners? Are they making consistent decisions, or is there any bias shown?

☐ Data – is this kept safe and secure? Does it comply with the data protection and confidentiality requirements?

☐ Decisions – do all assessors and internal quality assurers' decisions meet the relevant requirements, regulations and standards? Are accurate, dated and legible records maintained?

☐ Evaluation – does evaluation take place, i.e. can learners give feedback? Is this documented and acted upon?

☐ Induction – are all learners adequately inducted? Are records maintained?

☐ Internal quality assurer expertise, knowledge and competence, whether new, experienced, qualified or working towards an IQA qualification (and being countersigned if relevant) – staff should have appropriate job descriptions and development plans. Check CVs and original certificates. Check CPD records and how staff maintain their knowledge and competence.

☐ IQA practice – is there an adequate and up-to-date IQA rationale and strategy? Are all IQA records up to date and do they meet awarding organisation requirements? Does the internal quality assurer give support and developmental feedback to assessors? Do various IQA activities take place throughout the learner journey as well as at the end (interim and summative)?

☐ Language barriers – are these taken into consideration, e.g. English as a second language?

☐ Learning support – are learners given adequate and appropriate support as necessary? How has it been identified?

☐ Locations and allocations of learners and assessors, internal quality assurers – are staff accessible and can they be located? Are the allocations fair, i.e.

(Continued)

does one assessor have more learners than others for no particular reason? If there is more than one internal quality assurer, are they allocated fairly to their assessors? Do staff from different locations have the opportunity to standardise their practice?

☐ *Logo – is the centre using the awarding organisation's logo without permission?*

☐ *Malpractice – do you suspect something is seriously wrong, for example, signatures not matching, assessors signing off units which are incomplete, IQA activities being documented but not taking place? If so, you must contact the AO.*

☐ *Meetings – are these taking place regularly and minutes distributed to all? If an assessor misses a meeting, how are they updated?*

☐ *Policies and procedures – are these current and relevant, for example, appeals, health and safety, equality and diversity?*

☐ *Pressure – do internal quality assurers or assessors feel under pressure to pass learners who are borderline, perhaps due to funding, targets or employer expectations? Do you feel under pressure, i.e. not having enough time to carry out your role effectively to make valid decisions? If so, you may have to arrange a further visit via your awarding organisation. Never feel pressured because your centre is wanting direct claim status or rushing you to make a decision.*

☐ *Qualification criteria or standards – are these new or have they recently changed? Are all staff aware of the changes?*

☐ *Records – you could compare assessment records across all assessors. Are some more detailed than others? If so, this is a standardisation issue. Are records manual or electronic? How does the centre authenticate them? Do dates on tracking sheets agree with those on feedback records? Are there any causes for concern with electronic records, for example, access to learner work and assessment records, use of visual recordings of observations? Are all records kept confidential, safe and secure (usually for three years)?*

☐ *Registration – are all learners registered with the awarding organisation and how soon after they commenced does this take place? (You should not sample from a learner who is not registered with your AO.) If there is a long time gap, is this in breach of awarding organisation guidelines?*

☐ *Reviews of progress – do learners have the opportunity for regular reviews of their progress? Are they documented?*

☐ *Risk rating – what was the previous risk rating? If it was medium or high, has the centre addressed the issues? Can they now be on a lower rating and have their direct claim status back?*

☐ *Risks – are any assessors experiencing problems or have concerns (e.g. access to their learners for assessment purposes, learner issues, plagiarism, cheating, etc.)? Risk in this context relates to assessment risks, not health and safety risks.*

☐   *Sampling – have you sampled work from all assessors and all locations which has been both internally quality assured and not? Have you observed internal quality assurer feedback to assessors, as well as trainer and assessor practice (if applicable)? Have you spoken to learners and others, e.g. witnesses? Has the centre prevented you from sampling anything you have requested? If so, can they adequately explain why?*

☐   *Staff – is there a high turnover of staff? If so, why? Are learners allocated to assessors in a fair way? Do some staff have higher caseloads of learners than others, resulting in pressure to get them through?*

☐   *Standardisation – is there a plan and are activities taking place regularly with records maintained?*

☐   *Time wasting – do centre staff keep you talking or offer to take you out for lunch? If so, they could be trying to distract you or waste your time for some reason.*

☐   *Training – do learners receive adequate training prior to assessment? If the course is formally delivered is there an adequate scheme of work and appropriate session plans? Do staff have relevant training and development opportunities to keep up to date with developments in their subject area?*

☐   *Trends – are most learners making the same mistakes? If so, it could be that the assessor has misinterpreted something. You may have to widen your sample if you find a trend.*

☐   *Type of qualification being assessed – are there any problem areas or units that could be misunderstood?*

☐   *Witnesses – are the testimonies reliable? Do the witnesses really exist and are they competent to make a decision?*

The above checklist might make you feel a little daunted as to all the aspects you need to be aware of when sampling a centre's practice. However, once you get a feel for how your centre operates, you will soon get to know what to look for. If you have a gut feeling that something is not right, you will need to delve deeper to find it. You can't make any decisions based on instinct; they must be based on what you have sampled and seen.

Reviewing all the assessment and IQA documentation will help you gain a clear picture of how the centre functions. If witnesses are used, you should contact a sample of them to confirm their authenticity and that they understand what their role entails. If a qualification relies heavily on the use of witness testimonies as evidence, then you will need to ascertain how the centre supports them. You could talk to a sample of witnesses and document the responses using a checklist as in Table 4.9.

You need to ensure that all supporting assessment and IQA tracking records are seen for the learners' work you have sampled. Dates on assessment and IQA records must agree with those on the tracking records. If they don't, you will need to ascertain why – it could simply be an error or there could be a problem which will need following up. You should be able to follow a clear audit trail from when a learner commences to when they complete or leave.

You might come across something that causes you concern; if so, you will need to tact-fully discuss it with the centre staff. As a result, you may need to adapt your visit plan to deal with the situation. This might include amending what you will sample, observing more staff and/or talking to more learners. If this is the case, make sure you document in your report the reasons why you did not follow your original visit plan. Your awarding organisa-tion might want to know why you didn't do something that you had originally planned to. It could be that a learner you had asked to interview is away at the time of your visit. Ideally, the centre should have informed you of this in advance to enable you to choose another learner. If the centre substitutes a learner without informing you, you will need to know why. It could be a genuine situation such as the learner is ill, or it might be that the centre does not want you to talk to a particular learner for some reason. If this happens, make sure you sample the assessment and IQA records for the learner. This will confirm they do exist and that progress is being made.

You need to get a feel for how your centres operate and how you can trust what they tell you. As you only sample aspects of a centre's operation, you are likely to miss some things. Therefore, getting to know the staff, how they operate and encouraging them to commu-nicate with you regularly will help you understand if they are doing something intentionally or not.

Agreeing action and improvement/development points

Even if a centre is low risk, you might have identified some areas which require action, improvement or development. This could be due to the way they complete their records, perhaps a policy which needs updating or a handbook which is out of date. You should agree SMART action points with the centre staff and clearly document these in your report. SMART stands for Specific, Measurable, Achievable, Relevant and Timebound.

Action points

An action point is something specific, which relates to the operation of the centre or the delivery, assessment and internal quality assurance of a qualification. An action point can only be given when something does not meet the relevant requirements, regulations and standards. It cannot be given just because you want the centre to do something in a particular way. You must always be able to substantiate your action points by showing the centre the aspects to which they apply in a relevant document. Action points should have a target date; they are enforceable and should be monitored.

Example

Michael, the EQA, had asked to see the records of standardisation activities during his visit, and the centre could not supply them. They said they held meetings but did not keep records. The SMART action point on his report therefore read: **The internal quality assurer must hold standardisation meetings with all assessors to ensure all units of the qualification are standardised during a** yearly cycle. A plan must be created and e-mailed to the EQA within one month. Records of the activities must be e-mailed to the EQA within two weeks of the activity taking place. *Michael made a note to himself to check that the centre had sent the plan to him in a month's time.*

Improvement and development points

An improvement point can be given to a centre to help and support them to improve a particular aspect. A development point can be given to a centre to help develop their practice in general. Some awarding organisations use the term *improvement*, whereas others use the term *development*. Improvement and development points are recommendations and are not enforceable like action points are.

Example

Pauline, the EQA, noticed the centre was completing two separate observation forms, when one would be adequate. She advised the centre how to do this, so that they could still meet the qualification requirements. The SMART improvement point therefore read: The assessor will consider amalgamating the two current observation forms within one month, to create one form which will still meet the requirements of the qualification.

Awarding organisation report

Your awarding organisation will provide you with a report template which should be completed during or shortly after you carry out the sampling and monitoring activities. The report might differ depending upon the activity carried out, e.g. a visit or a remote monitoring activity. It will probably be electronic, allowing you to key in text into various areas but not allow you to change the questions. You should answer all the questions objectively, based on fact and not on opinions. You should check for spelling and grammar errors to ensure your report is completed professionally. Don't get so tied up with completing the form that you don't allow enough time to carry out the activities on your plan or to give feedback to the centre staff. Report forms will differ between awarding organisations. You will need to access and fully understand the requirements and the questions being asked on the report, prior to completing it. You should have been given training regarding how to complete the report, the amount of detail to be added, and when and how a copy will be given to the centre.

The report might be in sections to cover the following aspects:

- centre name, number and contact details
- EQA name and contact details
- date of activity and what was carried out

- locations and assessment sites visited

- staff, learners and witnesses who were met and/or interviewed

- what was sampled, e.g. learners' work, assessment and IQA records

- what was observed, e.g. assessors giving feedback to learners or IQAs giving feedback to assessors

- details of management and quality systems, policies and procedures

- details of resources and equipment

- appeals, complaints and disputes

- CPD, qualifications and experience of staff

- records such as minutes of meetings and standardisation activities

- feedback given to the centre

- actions and improvement points as required.

You must be able to back up any statements in your report, in case there is a query, a complaint or an appeal by the centre. The report should be passed on to the awarding organisation within their timescales. Some awarding organisations will forward it to the centre; others will request you to do this. Make sure you keep your own copy safe and use a filing system that enables you to access your records easily, yet keeps them confidential. If you have also completed any of your own documents, for example, observation checklists, make sure you file these safely too. If you have completed them manually, you could always scan or take a photo of them to save electronically.

After the monitoring activity, you should:

- ensure your report has been completed correctly and forwarded to the awarding organisation within the required timescale, and inform the centre how and when they will receive their copy

- ensure all your records are up to date, e.g. your centre contact record

- keep in regular touch with the centre, particularly if there are action points

- keep in regular touch with the awarding organisation as necessary

- keep track of centre activity, e.g. registrations and certifications; you might be able to do this via the awarding organisation's online system

- evaluate your practice to identify how you can improve.

Disputes and appeals

Most centres comply with all the relevant requirements, regulations and standards, however you might come across a centre that breaches these. This could be due to ignorance, or it might be deliberate. Sadly, some centre staff feel pressured due to targets and funding, and might commit fraud, such as completing some of a learner's work for them. If you find this during a visit or a remote monitoring activity, you should contact your awarding organisation and discuss it with them first. Whilst you can recommend a sanction, the awarding organisation has the final say. They can access the previous track record of the centre to see if similar problems have occurred in the past.

The staff might dispute your decision. Therefore it's crucial that you keep full and accurate records of everything you find, not only by completing the awarding organisation's report, but also in your own notes. If a centre does dispute your decision, the awarding organisation might ask another external quality assurer to carry out a sample from the centre to see if there is a trend emerging. Don't take anything personally; it's the situation not you that they will be disputing.

Another type of dispute might be within the centre, and as their EQA you could become involved to help them resolve it.

Example

Jeremiah, the EQA, was talking to a learner, Adam, during a visit to a centre. Adam stated he had failed an assignment because he felt the assessor didn't like him. Adam had complained to the internal quality assurer but had not received any feedback. Jeremiah looked at the assignment and agreed the assessor was right, but that he had not clearly explained to Adam what the requirements of the assignment were. Jeremiah asked to see the centre's appeals procedure, which could not be located. Jeremiah therefore gave the centre an action point to ensure the procedure was updated immediately and to inform all learners how they could access it.

You should always check that the centre has an up-to-date appeals policy and that all staff and learners are familiar with it. It might be included in a learner handbook or a centre manual, or be accessible via the centre's intranet or website. When talking to learners, this is a good time to ask them if they know what to do if they disagreed with their assessor's decision. You should also ask a centre if they have had any appeals and if so, look at the records. It might be that one particular assessor is receiving far more appeals than others. If so, you would need to find out why.

Dealing with bad practice

Hopefully, you will only identify points for action or improvement when carrying out a sample at a centre. If you are finding a trend, for example, all assessors misinterpreting the same thing, you could involve the IQA in a proactive way. You might ask them to sample something with you, to see if they notice what you have found. This way, they can take ownership of the problem and discuss with you what can be done.

If you find something that a person could improve upon, don't be critical but state the facts based upon what you have seen. After an observation, you could ask the person to reflect upon their performance before you give feedback. That way, they might realise any mistakes before you have to point them out. You can then suggest ways to improve. You might not find anything wrong, in which case you still need to give feedback, which will be positive and confirm that what they are doing is right. Always allow time for them to clarify anything you have said and to ask questions. Don't interrupt them when they are speaking and avoid jumping to any conclusions. Use eye contact, observe body language and listen carefully to what they are saying. Show that you are a good listener by nodding your head and repeating key points.

Example

Josh, the EQA, carried out an observation of Donna, the assessor, with one of her learners in the workplace. He used an observation checklist to note what he had seen and to enable him to give constructive feedback. He noticed Donna was really detailed at giving developmental feedback, but forgot to state which aspects her learner had achieved. He asked her afterwards if there was anything she would do differently, and she did state this point. As the learner was still available, Donna went to explain to him what he had achieved.

If you are carrying out a visit to a centre, you could give ongoing feedback, perhaps when you identify good practice, and then give further feedback at the end. Some centres might prefer you to give feedback to one named person, perhaps a manager or an internal quality assurer. Others might be happy for you to feedback to the full team. Always make sure you allow plenty of time for feedback and questioning.

Activity

Think about the last time you gave feedback to centre staff. Were they receptive to what you had to say, or did they become defensive for any reason? Would you change the way you give feedback in the future based on this? Did you allow enough time for them to ask questions, and if so, were you able to correctly answer them?

You need to be tactful when giving feedback, particularly if you have identified any concerns, as some staff might take things personally and become defensive. Although you will be on your own at the centre, never allow anyone to persuade you to change your decision or intimidate you with statements such as staff losing their jobs as a result of your findings. You must remain objective and do everything according to the relevant requirements, regulations and standards. You can always telephone the awarding organisation for advice, or simply say that you have to leave if you feel threatened.

Find out from your awarding organisation what you should do if a centre's practice does not fully meet the relevant requirements, regulations and standards. Find out what to do if you are faced with a difficult or threatening situation. Perhaps there is a lone workers' policy you should become familiar with. Your awarding organisation should help and support you. However, they might not always be available, for example, if you are working during the evening or a weekend. It's therefore important to know what to do in the event of a real problem at a centre.

Record keeping

Records must be maintained to satisfy the awarding organisation's information management systems, and relevant regulatory bodies. It's important that EQA reports, either manual or electronic, show a full audit trail of what has been monitored, sampled and when. If records are not kept, there is no proof the EQA process took place. This includes your own notes and checklists as well as the reports you complete for the awarding organisation. Maintaining records will usually be for a set period of time, for example, three years. Records should always be accurate, detailed, dated and legible.

There may be a standardised approach to completing the records and reports for an AO, for example, the amount of detail which must be written, whether the records should be completed manually or electronically, whether they should be written in the first or third person, and how they are completed, i.e. written, word processed or online. The awarding organisation should invite you to regular standardisation activities with other external quality assurers. This is an opportunity to see how they complete their records and carry out their role.

Activity

Contact another external quality assurer in the same subject area as yourself; your AO should be able to help you locate one. Find out what records they keep and how they store them. Ask them for any hints and tips they have regarding record keeping and completing the various EQA documents and reports.

When completing any records, if signatures are required, these should be obtained as soon as possible after the event, if they cannot be signed on the day. Any signatures added later should have the date they were added, rather than the date the form was originally completed. If you are completing documents electronically, you will need to find out if an e-mail address, scanned or electronic signature is required or not. This might be acceptable providing the identity of the person has been confirmed and a record kept of the original signature.

All records should be kept secure and should only be accessible by relevant staff. You also need to ensure you comply with organisational and statutory guidelines such as the Data Protection Act 1998 and the Freedom of Information Act 2000. See Chapter 1 for further details.

Managing information

Records of all communication and activities carried out with a centre should always be kept up to date. This will include who you have met, who you have interviewed and which learners' work you have sampled. You should monitor all staff over a period of time and should not assume that experienced staff will be performing adequately. It is easy for staff to become complacent; however, when standards change new practices may have to be implemented by the centre to ensure everyone is aware of the changes.

Records you might keep include:

- awarding organisation reports
- records of contact and communications with centres
- EQA visit and sample plans sent to centres
- observation records and checklists
- interview records, for example, with learners and witnesses.

Example

Anka is an EQA who works from home. She doesn't have a dedicated office, but has a computer in the corner of her living room. She has a file for each of her centres which she keeps in a cupboard. Whilst talking to another EQA at a recent meeting, she was told she must obtain a lockable, fireproof cabinet in which to keep her records. She also realised her electronic files could be accessed by her family, who also use her computer. She therefore ordered a small lockable metal filing cabinet for the manual files and set up a password for the computer files.

You must never trust your memory and you should document all communication with your centres between visits, as well as keep records of what was carried out during an approval or an advisory visit. Always take a backup copy of electronic documents. It would be dreadful if you had been completing a report online, only to find the internet connection had stopped before you had saved your work.

Reasons for keeping records include (in alphabetical order):

- for standardisation and quality assurance purposes
- in case of a complaint or appeal against your decision

- in case of malpractice in a centre, e.g. plagiarism or fraud

- to keep track of communications with centres

- to monitor qualification and staff activity within a centre

- to satisfy the awarding organisation's and regulatory requirements, i.e. you must be able to show a valid audit trail for all your activities.

Remember that records should always be accurate and based on what you have seen. They should be legible and kept safe, secure and confidential. Most records can be kept manually or electronically. Backing up your data and records is important, particularly if you have electronic records, for example, in the case of power failures.

Confidentiality

Confidentiality should always be maintained regarding all information and data you use. You might work from home, in which case you should ensure all your records and data are safe and secure. Using your car boot for records is not a good idea. You also have to ensure your centres are following data protection and confidentiality requirements for their records.

Extension activity

Find out what the policies and procedures are for your awarding organisation regarding confidentiality, data protection and record keeping. They might be accessible via the AO's website. Make sure you are familiar with them all. If you have any concerns, make contact with the AO to clarify any points.

Evaluating practice

It's useful to evaluate your own practice, and obtaining the views of others will greatly assist you when reflecting upon your role. Your awarding organisation might ask for feedback from your centres regarding how you have performed. This information might be used during their appraisal or review process with you. You could also ask for feedback from your centres to help you evaluate the service you are giving them. This could be done at the end of a visit, however if the visit has not gone well, you won't want to get into a detailed discussion.

Feedback from staff in centres might impact upon your role by enlightening you to other aspects, for example, how you gave advice, guidance and feedback. Don't always think that the feedback you receive will be negative. Often centres are happy with the service they receive as it can confirm they are doing things right, or that they could consider doing something differently.

Always make sure you do something with the feedback you receive which will lead to an improvement in your practice.

Why evaluate external quality assurance practice?

You should evaluate the full EQA process for each centre you are responsible for. This includes communication before, during and after a visit or a remote monitoring activity.

This will ensure:

- a professional service is given to centres

- the process is fair to all

- you are meeting relevant requirements, regulations and standards

- you can learn from any incidents

- you can improve your practice.

It is important to give a good service to your centres, and to maintain and improve on this where possible. However, the credibility of the qualification, reputation of the award-ing organisation and the quality of the service you provide must never be compromised. A compromise might occur due to being too friendly with centre staff rather than remaining objective and professional.

Example

Alex was partaking in the annual review of all external quality assurers. His lead EQA, Jim, had received feedback from all six of the centres Alex has been monitoring and used this as a basis for the meeting. Feedback was very positive regarding the advice and support Alex had been giving, both during visits and in between. All the centres were impressed with his knowledge and professional-ism. One centre in particular had commented how helpful Alex was when they had a particular problem. However, one centre felt he had not spent enough time answering their questions. Alex accepted this and agreed he would allow more time for centre staff to ask questions, perhaps at the beginning of the visit as well as at the end.

Whether you partake in an appraisal or a review process or not, you should self-evaluate aspects to help you improve. This could include:

- How effective was my communication with the centre staff?

- How do I keep adequate, accessible and secure records?

- Did I carry out everything I had planned to do during a visit or a remote monitoring activity? If not, why was this?

- Did I deal with any awkward situations effectively?

- Do I allow enough time for questions from centre staff?

- Did I give constructive and developmental feedback?

- Did I complete the report correctly and agree relevant action and improvement points?

- Did I ask for feedback from the centre regarding my performance?

- Do I standardise my practice with other external quality assurers?

- Do I keep up to date with changes and developments regarding my subject area?

Activity

Answer the questions in the bullet list above. If the question requires a yes or no response, ask yourself why this is and what you could do differently. What other questions could you ask yourself to help you review your practice?

Please see Chapter 2 for further information regarding evaluating practice and self-evaluation.

Standardisation of practice

The support you give to one centre should be similar to what you give to another. You therefore need to ensure a standardised quality service is given to all your centres. This should be comparable to that given by other external quality assurers to their centres. You must not show any favouritism or do things which are not *by the book*, i.e. you must follow all written guidelines and regulations. You must not ask a centre to do something which is not a requirement just because you want them to. You need to be fully aware of your awarding organisation's policies, procedures and practices, and follow all relevant legal and other regulatory requirements. The awarding organisation should provide you with written guidance which you should become familiar with. You should also attend any standardisation activities which are provided by your awarding organisation. This is an opportunity to meet other EQAs and discuss and compare how they carry out their role.

Activity

How can you standardise the way you perform your role with that of other external quality assurers? Find out from your AO when the next standardisation meeting is. Prepare a few questions that you might like to ask when there.

You could standardise your practice in the following ways (in alphabetical order):

- carrying out peer EQA observations (with the approval of the AO and centres)

- communicating with other external quality assurers as to how they carry out their role and support their centres in your subject area

- comparing decisions from other EQAs, i.e. by reviewing anonymised documents and reports as part of an awarding organisation's standardisation event

- keeping up to date with changes to qualification criteria and standards to ensure you interpret them in the same way as others

- requesting an accompanied visit from another external quality assurer to go with you to one of your centres.

Example

Bronia has been working as an external quality assurer for several years. A new external quality assurer, Hebe, has just commenced and Bronia has been asked if Hebe could accompany her on a visit. Bronia is happy for Hebe to do this. It will help both of them standardise their practice and help reassure Hebe that she is not working in isolation from other EQAs.

Reflective practice

Reflective practice is an analysis of your actions which should lead to an improvement in practice. It can be written down, or just thought through. It is useful to reflect on your own practice, as well as on the feedback you receive from centres and the awarding organisation to help you improve and develop.

Part of reflection is about knowing what you need to change. If you are not aware of something that needs changing, you will continue as you are until something serious occurs. You may realise you need further training or support in some areas therefore partaking in relevant CPD should help. Please see Chapter 2 for further information regarding reflective practice.

As a professional, you need to continually update your skills and knowledge. This knowledge relates not only to your subject area, but also your practice as an external quality assurer and your knowledge of assessment and internal quality assurance.

CPD can be formal or informal, planned well in advance or opportunistic, but it should have a real impact upon your role and lead to an improvement in your practice. CPD is more than just attending events; it is also about using critical reflection regarding your experiences, which results in your development and leads to a positive improvement in your practice.

Planning and maintaining continuing professional development

In education and training, changes often occur, and you will need to keep up to date with these. They include qualification revisions, standards updates, changes to regulations and relevant assessment and quality assurance policies and procedures. Keeping up to date is all part of your continuing professional development.

Feedback from others and your own reflections will help you realise what CPD you need to undertake. You could shadow colleagues to observe how they carry out their EQA role, join professional associations, and carry out internet research regarding your specialist subject, assessment and IQA practice.

Examples of opportunities for CPD include:

- attending events and training programmes
- attending meetings
- e-learning and online activities
- evaluating feedback from peers, assessors and others
- formally reflecting on experiences and documenting how they have improved practice
- improving own skills such as English, maths and ICT
- keeping up to date with relevant legislation
- membership of professional associations or committees
- observing colleagues
- reading textbooks and journals
- researching developments or changes to your subject
- secondments
- self-reflection
- shadowing colleagues
- standardisation activities
- studying for relevant qualifications
- subscribing to and reading relevant journals and websites
- using social media to follow/inform others of relevant and current information
- voluntary work
- work experience placements
- writing or reviewing books and articles.

Records must always be kept of any CPD activities undertaken which might need to be shown to your awarding organisation or regulatory bodies if requested. Please see Table 2.12 in Chapter 2 for an example of a CPD record, along with a list of useful websites for keeping up to date.

Reflecting upon your practice, taking account of feedback and maintaining your CPD will all contribute to becoming a more effective external quality assurer.

Look at the previous bullet list and consider what CPD activities you could carry out to improve your role as an external quality assurer. This relates not only to your role, but also to the subject you will externally quality assure, and any guidelines and regulations you need to be familiar with. If you are currently carrying out an external quality assurance role, reflect upon your last visit to a centre to consider what you would change or improve and why.

Summary

Carrying out your role as an EQA in an organised and professional manner will help ensure a centre's assessment and IQA decisions are accurate, consistent, valid and reliable.

You might like to carry out further research by accessing the books and websites listed at the end of this chapter.

This chapter has covered the following topics:

- External quality assurance planning

- External quality assurance activities

- Making decisions

- Providing feedback to centre staff

- Record keeping

- Evaluating practice

References and further information

Ofqual (2015) *General Conditions of Recognition*. Coventry: Ofqual.

Pontin, K. (2012) *Practical Guide to Quality Assurance*. London: City & Guilds.

Read, H. (2012) *The Best Quality Assurer's Guide*. Bideford: Read On Publications Ltd.

Roffey-Barentsen, J. and Malthouse, R. (2009) *Reflective Practice in Education and Training* (2nd edition). London: Learning Matters.

Scales, P., Pickering, J., Senior, L., Headley, K., Garner, P. and Boulton, H. (2011) *Continuing Professional Development in the Lifelong Learning Sector*. Maidenhead: OU Press.

Wilson, L. (2012) *Practical Teaching: A Guide to Assessment and Quality Assurance*. Hampshire: Cengage Learning.

Wood, J. and Dickinson, J. (2011) *Quality Assurance and Evaluation in the Lifelong Learning Sector*. Exeter: Learning Matters.

Websites

Evaluation – www.businessballs.com/kirkpatricklearningevaluationmodel.htm

Plagiarism – www.plagiarism.org and www.plagiarismadvice.org

Reflective practice – http://www.learningandteaching.info/learning/reflecti.htm

5 PLANNING, ALLOCATING AND MONITORING THE WORK OF OTHERS

Introduction

If you lead a team of assessors, internal or external quality assurers, you should be in a managerial role, be able to make decisions and take responsibility for others. You will need to plan, allocate and monitor your work and the work of those in your team within the context of their roles.

This chapter will explore how to plan, allocate and monitor the work of others. Various leadership, management and communication theories are explored.

This chapter will cover the following topics:

- Producing and using a work plan
- Identifying and allocating responsibilities to team members
- Monitoring the progress of others and the quality of their work
- Communication skills
- Updating the work plan

Producing and using a work plan

If you are responsible for a team of assessors, internal or external quality assurers, you should produce and use a work plan. This is a visual reminder of what needs to be done and when. Using one will help you plan your objectives and the various activities required to meet them, by yourself or members of your team. An example objective could be: *to plan, monitor and review the internal quality assurance (IQA) process in accordance with organisational requirements.* Your work plan would then list certain activities such as: *create an IQA sampling plan.* See Table 5.1 on page 157 for an example work plan for an internal quality assurer who leads a team of other internal quality assurers and assessors (often referred to as a *lead internal quality assurer*). The shaded boxes are the months in which the activities will take place; the dates are then added when the activities have occurred.

When you are delegating tasks to team members, having a work plan will help ensure you allocate work fairly and effectively. Monitoring and reviewing the activities will ensure appropriate progress is being made. The tasks that your team members carry out should lead to an improvement in the quality and standards of the product or service being offered. The product could be a qualification or a programme of learning. The service could be the ways the staff support the teaching, learning and assessment process. Records should always be maintained of all activities carried out for audit purposes and for external inspections if required.

You might be new to leading a team, for example, if you have been an assessor or internal quality assurer as a member of a team and then promoted to lead that team. You might still carry out the same roles as before, but you will now have additional managerial roles as a team leader. You will need to appreciate that how you acted with your peers might need to change now that you are managing them. For example, you might not feel you want to be too personal and friendly with them. Therefore you will need to keep your working relationships on a professional basis.

Your aim as a manager of a team should be that your own work, and the work of others, is carried out efficiently and effectively. Your starting point for this will be your job description or role specification. This should state what you are expected to do when leading your team. If you don't have a job description which specifically covers this aspect, you could look at or work towards the requirements of a relevant unit in a management qualification.

Activity

Have a look at your job description or role specification to see what is stated regarding managing a team. If you don't have one, make a list of what you consider your management roles and responsibilities to be. If you are working towards a management qualification, have a look at the specification to see what you currently know and can do. Make a note of anything you are unsure of and then revisit your notes after reading this chapter.

As a team leader or manager you should be familiar with and committed to the organisation's purposes, values and goals. These are often defined as part of a vision and/or mission statement. A vision statement describes the organisation's aims and where it hopes to be in the future. A mission statement defines the organisation's business strategy, objectives and how it hopes to reach those objectives.

You will need to know what part you have to play in achieving them, as well as what your team members will need to know and do. The vision and mission of the organisation must be achievable. All staff should understand what they are and be committed to them. However, if they are unrealistic, then staff might not be motivated to achieve them. If the latter is the case, you will need to discuss issues and concerns with someone in authority, otherwise staff morale may become low which could lead to staff not performing their roles correctly.

| **Activity** |

Obtain and read your organisation's vision and mission statements. Analyse how they will impact upon your role and the roles of your team members. Think about how you will convey them to your team members to ensure they are all understood.

Having knowledge of how the vision and mission statements might impact upon your own role and the roles of others will help you when devising your work plan. If you currently carry out internal or external quality assurance activities as well as manage a team of staff, you might already be using some types of work plans such as:

- an observation plan

- a meeting and standardisation plan

- a sample plan and tracking sheet.

However, you might have other responsibilities that need to be planned for, or activities you need to delegate to others. When delegating, try not to impose the activities, but discuss and agree them to ensure you are utilising a person's strengths for the task. A work plan will help you prioritise and keep track of all objectives and the activities to meet them.

Information you might need to help create a work plan includes (in alphabetical order):

- details of your team members, what their experience, knowledge and skills are, and what they are expected to do and when

- financial information such as budgets

- information regarding what is being assessed and quality assured

- job description (or role specification) for yourself and others you are responsible for

- organisation vision and mission

- priorities, targets and expected success criteria

- relevant documentation, policies and procedures

- relevant requirements for accredited and endorsed qualifications (if applicable)

- resources: physical and human, and other aspects such as transport, and availability of resources

- the locations and contact details of your team members and other relevant staff.

If you are not familiar with any of the items in the bullet list, you will need to find them out prior to working with your team.

Using the *who, what, when, where, why* and *how* (WWWWWH) approach will help you create your work plans. All tasks and activities you set for yourself and your team should have objectives which are SMART, i.e. Specific, Measurable, Achievable, Relevant, and Time bound.

Example

Harry has used the WWWWWH approach to help him create his first work plan. He has a team of four internal quality assurers who are each allocated five assessors. Here are his notes:

> *Who – the names of the four IQAs and 20 assessors, along with their contact details*
>
> *What – the activities which are to be carried out*
>
> *When – when the activities will take place*
>
> *Where – where the activities will take place*
>
> *Why – the reason for the activities*
>
> *How – how the activities will be conducted*

Based on these notes, Harry's SMART objective is: To plan, monitor and review the IQA process in accordance with organisational requirements. His first activity is: To produce an IQA sampling plan. See Table 5.1 for an example of a work plan.

A work plan can be produced and updated electronically, for example, as a spreadsheet, and then e-mailed to all staff in your team. Alternatively, it can be printed and displayed on a noticeboard as a visual reminder. If it is electronic, it's best to save any updates as a different version number to allow the original to remain accessible. Alternative styles of work plan could be used, for example, wall planners, or templates such as Gantt charts which are available free via an internet search or as part of some computer programs. A Gantt chart is so called as it is named after Henry Gantt (1861–1919) and is a type of bar chart used to illustrate a project schedule.

Activity

Create a work plan for a particular objective which is part of your area of responsibility. For example, To plan, monitor and review the IQA process in accordance with organisational requirements as in Table 5.1. What activities will need to be carried out and what will the target months be? What activities could you delegate to your team members and why?

Table 5.1 Example work plan for a lead internal quality assurer

Objective:	To plan, monitor and review the IQA process in accordance with organisational requirements (Customer Service qualifications)												
IQA name:	H. Rahl												
Activities & month:	Jan	Feb	Mar	Apr	May	Jun	Jul	Aug	Sep	Oct	Nov	Dec	Comments
Produce IQA sampling plan for self, ensure other IQAs complete theirs	5th												
Produce IQA observation plan, ensure other IQAs complete theirs	5th												
Plan team meeting dates and delegate the role of the chair on a rota basis	12th												
Plan standardisation activities and dates		18th											
Prepare for external quality assurance visit			�										Inform all staff and hold an additional meeting before and after
Meet with external quality assurer				▪									Reserve a room
Review assessment/IQA policies and procedures, including IQA rationale and strategies					▪								
Review assessment and IQA documentation						▪							
Carry out staff appraisals									▪	▪			
Produce report and statistics regarding appeals and complaints											▪		Could be done in December if necessary
Write annual IQA report for directors											▪		Distribute two weeks beforehand
Attend annual directors' meeting												▪	

Supporting your team

You might have staff who are employed full time, part time, are freelance, or work on a voluntary or peripatetic basis (i.e. working for several organisations). You will need to know who all your team members are and keep their contact details handy. There will be occasions when you might not all be able to get together at the same time for a meeting, therefore using electronic methods of communication can ensure you all stay in touch regularly.

To support your staff effectively, you will need an in-depth knowledge of all relevant policies and procedures, as well as any legislation you (and your team members) must follow.

Activity

Locate the relevant policies and procedures that underpin your job role and ensure you are familiar with them. Find out what legislation you need to follow (e.g. health and safety, equality and diversity, data protection, employment law, etc.). If you are unsure of anything, it could be that your team members are too. You will therefore need to discuss this with them to ensure everyone feels comfortable in their role, perhaps at your next team meeting.

You should support your team members in a proactive rather than a reactive way, encouraging them to talk to you when they need to. If they have a concern, it will need to be dealt with straightaway rather than it escalating into a major issue due to a lack of communication. However, there might be occasions when you have to deal with poor practice; in these cases you will need to ascertain all the facts and remain impartial with any decisions you make.

You might need to carry out coaching and/or mentoring activities with your team members, carry out staff appraisals, training needs analyses and/or countersign unqualified staff's decisions. Countersigning confirms that the decisions made by staff working towards an assessor or quality assurance qualification are correct. You might also be responsible for producing and/or updating staff and learner handbooks and other documentation such as assessment and IQA rationales, strategies and documents. If you haven't already read the preceding chapters regarding internal or external quality assurance, you will find lots of information to help with your knowledge.

Table 5.2 lists the skills required to help you perform your job role and to support your team. If there are any aspects you are uncertain of, you could research them further.

When planning activities for your team members to carry out, you will need to match their skills, qualities and locations against certain objectives and activities. You will need to know your staff well to be able to do this, or talk to them to find out their strengths and limitations. You might need to work within specific time or financial constraints such as deadlines or the lack of a suitable budget. Although all activities will still need to take place and be monitored, they could be carried out at a different time or location or by using technology. Your work plan can be updated and amended at any time to take into account any changes or unforeseen events which might occur.

Improving your own skills and those of your team members will help everyone perform their job roles to the best of their ability and contribute towards their continuing professional development (CPD). There might be instances when your team members have become demotivated or demoralised by circumstances beyond their control, for example, a recent spate of redundancies at the organisation or certain individual issues. Your own personal skills and qualities could help motivate and enthuse your team members.

Table 5.2 Example job role skills

Skills needed to perform your role and support your team include:	
• analysing	• mentoring
• appraising	• monitoring
• assertiveness	• motivating
• budgeting	• negotiating
• coaching	• planning
• communicating	• prioritising
• computing	• problem solving
• consulting	• providing feedback
• data analysing	• questioning
• decision making	• reading and writing
• delegating	• record keeping
• leading	• reviewing
• listening	• setting SMART objectives
• managing: conflict, people, problems, projects, stress, time	• speaking
	• supporting others

Activity

Make a list of the personal skills and qualities you feel you currently possess and rate yourself 1, 2, 3, 4 or 5 (1 being low and 5 being high). Compare these to those listed in Table 5.2. Now make a list of any skills and qualities you may need in the future and what you can do to achieve them. Whilst doing this, consider your roles, responsibilities and the management duties you are/will be performing.

Some people are naturally good managers, have a helpful and pleasant personality and are good at communicating with others. However, this doesn't come naturally to everyone. If you need to identify and improve the performance of others and monitor their progress, appropriate training will help you if you can get it. You should get to know your team members and realise that they may perceive your tone of voice and body language in a different way to that which you intended. Communication is important and it should always be carried out in a professional manner. You might also need to appreciate that your personal values and beliefs might be different to those of your team members, and not impose yours upon them. You might have to deal with staff complaints, appeals and grievances, either directly or indirectly. If this is the case, you must remain impartial and maintain confidentiality throughout the investigation process. Communication skills will be covered later in this chapter.

Personality styles and preferences

Understanding a little about your personality style or preference will help you develop your personal qualities and skills. There are many models that can be used to ascertain personality styles and preferences; the following are a few examples, which you might like to research further.

In the 1970s, a test was devised by cardiologists Friedman and Rosenman to identify patterns of behaviour considered to be a risk factor for coronary heart disease. This placed people as *Type A* or *Type B* and this is still used today to analyse personality types. Type A individuals can be described as impatient, excessively time-conscious, insecure about their status, highly competitive, hostile and aggressive, and incapable of relaxation. They are often high-achieving workaholics who multitask, drive themselves with deadlines and are unhappy about the smallest of delays. Due to these characteristics, Type A individuals are often described as *stress junkies*. Type B individuals, in contrast, are described as patient, relaxed and easy-going. There is also a Type AB mixed profile for people who cannot be clearly categorised. Knowing which type you are could help you change if necessary, for example, becoming less stressed.

Other personality styles' tests include the Myers Briggs' Type Indicator (MBTI) and Kiersey's Temperament Sorter, which are based on the work of Carl Jung (1875–1961). Both involve taking a personality questionnaire. You might like to carry out a personality style test yourself and encourage your team members to do so too. However, the tests should always adhere to a code of ethics, and guidance for this should form part of the test.

Reviewing your personality style should help you see aspects you might need to change, improve or develop, and can help you become more effective as a leader or manager. Knowing people are of different types should help you realise how individuals act and react in different situations.

Taking personal responsibility for your own actions, supporting your team members and not apportioning blame should lead to a healthy working atmosphere for all concerned. This in turn should lead to an improved service for the learners and an improved reputation for the organisation.

Extension activity

Research a few different personality style questionnaires, such as Friedman and Rosenman's Type A and Type B personalities, Myers Briggs' Type Indicator (MBTI) and Kiersey's Temperament Sorter. Compare the differences and similarities. Decide if you would want your team to complete a questionnaire; if so, which one would it be and why?

the weaknesses and threats or get rid of them altogether. This might be a long-term project and you may need to work with others to achieve it.

Once you get to know your team members and their strengths, you should feel confident at delegating various activities to them. However, never assume they will always be carried out. You will need to monitor what your staff are doing, perhaps by asking them to give you a written, electronic or verbal report at certain times.

Example

Raj has a team of six internal quality assurers who each have four assessors for the Health and Social Care qualification. One, Cameron, is new and working towards his IQA qualification. Raj could allocate any of the experienced staff to countersign Cameron's decisions. After he asked his staff to complete a SWOT analysis, he allocated Tom due to the fact his strengths included being patient, approachable and a qualified mentor. Raj felt these strengths would further support Cameron's development and help him achieve his IQA qualification quicker.

At some point, you might experience some hostility between your team members, for example, if they feel they should or should not be carrying out various activities. Getting people to work as a team can be problematic, particularly if there is a high staff turnover in your organisation. If this is the case, you would need to find out why staff are leaving and meet with relevant managers to try and do something about it. Gaining feedback from the staff who have left could help you ascertain the reasons why.

Team working

A team is a collection of individuals and each person will have different ideas and ways of performing. Team-building activities can be good fun and lead to staff having a better working relationship with each other. However, some staff may feel they are a waste of time, or there might not be a suitable budget available for them. If you expect your team members to work together collaboratively on certain activities or projects, you might like to familiarise yourself with Tuckman's Group Formation theory of *forming, storming, norming* and *performing*. This was formulated in 1965 and amended in the mid-seventies to add *adjourning* (also known as *mourning* or *deforming*). There have been various adaptations of this theory over time, which you might like to research further.

Forming – this is the *getting to know you* and *what shall we do* stage. Individuals may be anxious and need to know the boundaries and code of conduct within which the team will work.

Storming – this is the *it can't be done* stage. It's where conflict can arise, rebellion against the leader can happen and disagreements may take place.

Norming – this is the *it can be done* stage. This is where group cohesion takes place and norms are established. Mutual support is offered, views are exchanged and the group co-operates to work towards the task.

Performing – this is the *we are doing it* stage. Individuals feel safe enough to express opinions and there is energy and enthusiasm towards achieving the task.

Adjourning – this is the *we will do it again* stage. The task is complete and the group separates. Members often leave the group with the desire to meet again or keep in touch.

Being aware of the stages that groups go through, and informing your team members of each, should help you all see why things happen the way they do. Depending upon the activity, the stages might happen over a short- or long-term time frame. You might even see all the stages occurring during one meeting where the team is new and has a tight deadline for an activity. Alternatively, you might have a team that gets stuck at one of the stages and you will need to intervene to move them on.

Activity

Plan an activity that you could carry out with your team, for example, creating or revising a document or template. Carry out the activity and watch how the team develops through Tuckman's stages. How many stages did it go through and was the activity achieved within the time allocated? Did you have any staff members who were disruptive or worked well with some team members and not others? If so, what could you do next time to ensure everyone is performing on task?

There are many other group formation theories available. You could try one of the following with your team or research others.

Coverdale (1977) stated the essence of team working is that individuals have their own preferred ways of achieving a task, but that in a team they need to decide on one way of achieving this. In a team, three overlapping and interacting circles of needs have to be focused upon at all times. The *task needs*, the *team needs* and the *individual needs*.

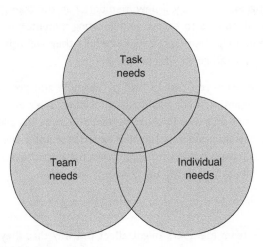

Figure 5.1 Team-working needs

When setting activities for your team members to carry out in groups, consider the following:

To achieve the task ensure:

- a SMART objective or target is stated
- responsibilities are defined
- working conditions are suitable
- supervision is available.

To build and maintain the team ensure:

- the size of the team is suitable for the task
- health and safety factors are considered
- consultation takes place
- discipline and order are maintained.

To develop the individual ensure:

- responsibilities are defined
- grievances are dealt with
- praise is given
- individuals feel safe and secure.

Individual personalities and the roles people take on when part of a group may impede the success or the achievement of the task. As a manager, you may need to make sure you supervise your team members' work carefully to keep all individuals focused.

You might need to agree some ground rules with your team members when working on group activities, for example, not checking mobile devices and respecting others' opinions to help the group progress effectively. If new members join the team, they should be made to feel welcome and introduced to everyone. Allocating a mentor to them, or acting as a mentor yourself, should help them settle in. Hopefully, they will not feel isolated as they have a named person they can go to with any questions.

As the manager of the team, you should lead by example and promote an environment based on respect. If a team member has made a mistake, don't directly blame them, but find out what went wrong and why. It could be that the same mistake might be made by others. You could use the situation as a learning experience and share your solutions with your team members to ensure it doesn't happen with them too. Your staff should feel comfortable to be able to talk to you about any concerns they have.

Team age ranges

The age range of your team members might affect the way that they work, or their attitudes towards work. The demographics of the country are continually changing. The *veteran* generation (aged 65 plus) may have been with the same employer for a long time and be thinking of retiring, have probably paid off their mortgage, have children who have left home and therefore have different priorities from the other generations.

The *baby boomers* (born 1946–65) might be working fewer hours and increasing their leisure pursuits, have grown-up children and a low mortgage. This generation will increase over the next few years and lead to a larger number of older than younger people in the workplace.

Generation X (born 1966–76) might be mid-career, have had several jobs, and perhaps experienced redundancy and unemployment along the way. They might have a large mortgage and a growing family.

Generation Y (born 1977–94) might be unemployed, be in training, be first or second jobbers, could still be living with their parents, have few responsibilities and possibly have large debts. They use technology a great deal and the line between work and social use can become blurred. They might have the desire to check work e-mails in their own time to keep on top of things.

Generation Z or the millennium generation (born 1995 onwards) have had lifelong access to technology, the internet and social media. Access to multimedia to such an extent can lead to a change in communication methods, which to other generations can look like a lack of social manners. Communication becomes via technology rather than face to face and can lead to poor spelling and grammar. Personal aspects often take priority over work due to the *immediate* and *switched on* lives they lead. This generation has been subjected to a fame culture through the many reality television shows and are often influenced by celebrities and fashion. However, an economic downturn may lead to a change in this generation's attitudes, for example, their concern for the environment, not being as indulgent as their parents were, and recycling and reusing products.

With these different generations come different aspirations, expectations, attitudes and values towards work. You might have team members from the different generations who have experienced these differences first hand. As a result, their attitude might be different towards their peers or indeed towards yourself as their manager.

Example

Martha is an assessor in her late 50s and extremely competent at assessing her subject of Hospitality and Catering. Her organisation wants all staff to use electronic assessment records via a mobile device, but Martha is apprehensive as she is not confident about using one. Her manager asks another assessor, Robbie, who is in his 20s to help her. Robbie is pleased to do this as he has been using mobile devices for many years. Martha was initially concerned and had even considered leaving, however she finds she gets on very well with Robbie. In return she shares a lot of her Hospitality and Catering knowledge with him which helps his understanding of his job role.

One of the greatest differences between the generations seems to be what they want from work. Generation X is enticed by freedom and independence and get on with their jobs without asking too many questions. Generation Y is more money and lifestyle oriented, focused on their own interests and used to 24-hour access to products and services. Generation Z has been brought up with technology and see it as embedded in, and integral to, their life. They have had the opportunity to use the internet, e-mail and various computer programs at school, and expect to use them at home and work. They embrace new technology with ease, expect instant access to information, use social net-working sites, and often don't have independent thought or retain information as it is so quick and easy to locate elsewhere. Self-development or self-gain is often part of their motivation and many are reward oriented. To them, their social life comes first. Baby boomers, in contrast, mainly use technology for convenience, for example, online banking and shopping, but it might not play a big part in their social or working lives.

Having an awareness of these differences might help you understand your team members' strengths and limitations. For example, recent graduates might be academically qualified but be lacking relevant experience. Older people, although very experienced, might not have the technological skills required. Being aware of this will help you appreciate the different aspirations, expectations, attitudes and values of the various age ranges within your team.

Extension activity

Research various management theories regarding teamwork, for example, John Adair (John Adair's 100 Greatest Ideas for Effective Leadership and Management, 2002), Meredith Belbin (Team Roles at Work, 2010) and Handy and Constable (The Making of Managers, 1988). Analyse their similarities and differences. Understanding more of the theories of why individuals act the way they do will help you plan the workload to the strengths of your team members.

Monitoring the progress of others and the quality of their work

You must regularly monitor the progress of your team members to ensure they are performing satisfactorily and meeting the required objectives. Your staff should be aware of what you will be monitoring and when. This can be planned formally by using your work plans.

When monitoring the progress of staff, you will be looking at the quality and consistency of their work, for example, how they complete various documents and the amount of detail they write. Records of all formal activities should always be maintained to show an audit trail of what has taken place. The records can also be used as a basis for improve-ment and standardisation activities. Records can be manual or electronic, and you will need to decide on how signatures should be used (or not) if it's the latter.

When monitoring the progress of your team members, you should talk to your staff regu-larly and give ongoing feedback. This should help:

- avoid potential problems

- boost motivation

- build a sense of ownership and responsibility

- build trust and respect

- confirm staff are doing their job correctly

- identify any training and development needs

- improve communication

- overcome resistance to change

- reduce potential conflict.

Activity

How will you monitor the work of your team, and what records will you maintain to show various activities have taken place? What will influence the way you do this, for example, relevant policies, procedures and legislation?

There will be internal and external policies, procedures and legislation which will need to be followed. For example, if you lead a team of external quality assurers your awarding organisation might require you to accompany each one to a centre visit at least once per year. You might consider this wrong if you or your staff are very experienced. Or you might consider it good practice in that everyone is being treated fairly. You will also need to know the limits of your own role and what you can and cannot do in the timescales you have.

Whatever methods you are using to monitor your team, you are ultimately ensuring they are being consistent at meeting their job role requirements and are correctly interpreting the requirements of the qualification or programme of learning (if applicable). The learner is the customer and needs to be treated fairly and ethically by everyone who is in contact with them. You also need to make sure everyone understands and complies with all internal and external requirements. This is particularly important if you are quality assuring accredited qualifications on behalf of an awarding organisation.

Although you are managing a team, you also need to ensure your own performance is meeting the required standards. If there are other internal or external quality assurers in the same subject area as yourself, you can monitor each other. If not, you could ask for feedback from others to help you improve. For example, it could be that you are performing your role adequately, but some staff feel you need to be more accessible when they have urgent questions.

You might have heard of the term *360 degree feedback*, which is a way of evaluating performance. It can contribute to staff appraisals and the self-evaluation process. It includes obtaining feedback from an employee's immediate work circle, e.g. their subordinates, peers, team members, supervisors and managers. It can also include feedback from external customers, for example, learners via questionnaires and surveys. The rationale for 360 degree feedback is that managers might not always fully understand the workload and contributions of their staff. Feedback from others can be as valuable as traditional hierarchical feedback from managers.

Activity

How do you seek the views of others? Do you have any evidence of obtaining feedback and acting upon it? You could consider devising and using a questionnaire to issue to your team members. The results might help you revise and improve the ways you monitor your team, and how you delegate certain activities.

When standards are not maintained

There are times when standards might not be maintained, for example, due to a lack of skills or knowledge by team members, or a lack of communication from managers. Identifying any issues early on can enable staff training and development to take place. The issues can then be used to standardise practice within the team to ensure it doesn't happen again. When allocating work to staff, you need to identify any potential issues as well as any individual skills shortages. This could mean you can't deploy someone as you would like to, or they could be placed in a difficult situation that they can't manage.

Example

Sharron regularly consults with her team members regarding how she will monitor their performance. She discusses with them what would happen if they don't meet the required standards. When performance does not meet these, it is usually because a member of the team needs further training. This is then arranged as necessary.

Monitoring your team members' performance and progress will help ensure everyone is receiving a good quality service. It should also identify any concerns that can be dealt with immediately, to alleviate any possible issues, appeals or complaints.

Unsatisfactory performance

Unsatisfactory performance could be due to many reasons, perhaps personal or professional. This could be an individual's fault, for example, not being honest with regard to their knowledge when they applied for the position. Or it could be the fault of the organisation by not conveying crucial information, for example, when a policy has changed. Recruiting and retaining staff who have the necessary skills, knowledge and experience will hopefully keep staff turnover low. If the organisation invests in training and ongoing support for team members, this should enable them to carry out their job role effectively. However, there may be times when an individual does not perform satisfactorily.

There should be a policy within your organisation for dealing with unsatisfactory performance, which might lead to disciplinary action if an individual does not conform. It might not be your role to deal with any issues, therefore you may need to liaise with the person whose role it is, perhaps from the human resources department. For example, if an individual's performance is not up to standard, an informal discussion could take place first to establish the reasons and agree any necessary action. If the action cannot be reasonably achieved due to individual circumstances, the staff member should be given the opportunity of support, such as further training or assistance. In more serious circumstances, they could be offered a reduction in their workload, some counselling or stress management.

Disciplinary action could occur if the individual's performance constitutes misconduct. There will be organisational procedures to follow and the process might be linked to performance appraisals and reviews. Formal records should always be maintained and relevant employment law followed.

Hopefully, your staff member will improve, make progress and meet the standards required. If you don't want the situation to escalate through the disciplinary procedure, you might be able to offer them a different job role or look at reducing their workload or hours. There are many reasons why someone might not be performing well and these should all be taken into consideration. You will need to liaise with relevant personnel in your organisation, such as those from the human resources department, to ensure any contracts and/or terms of employment/equality legislation have not been breached.

Unsatisfactory performance of an individual could reflect badly on you and your organisation. It is therefore important to identify any issues quickly, discover the causes and put an action plan in place to rectify the problem. If staff members are not performing adequately, there may be an impact upon others.

Activity

What do you consider to be poor performance and how would you deal with it, or who would you liaise with? What relevant policies and procedures are in place regarding unsatisfactory performance or disciplinaries?

Armstrong (2003) suggests the following guidelines for defining effective individual performance.

- Measures should relate to results, not efforts.

- The results must be within the individual's or team's control.

- Measures should be objective and observable.

- Data must be openly available for measurement.

- Existing measures should be used or adopted wherever possible.

Performance measures should always be discussed during staff appraisals and you should monitor the progress of individuals and teams towards them. There's no point making these measures unobtainable as you could be setting your team members up to fail. You are there to support your staff, not make things difficult for them.

Extension activity

How do you decide upon the standards of performance for the activities you expect your team members to carry out? What methods do you use to monitor the activities, and how effective do you think your methods are?

Communication skills

Communication is the key to effective management of your team members. The four skills of language are *speaking, listening, reading* and *writing*. Using these effectively in various ways should help with the achievement of the activities you expect your team members to carry out. Different methods of communication can be used depending upon the situation or person, for example, using the telephone if a staff member does not have access to e-mail.

Methods of communication include written and oral, for example:

- e-mail, texts or social networking – to quickly pass on information to the full team
- face to face – meetings or staff appraisals (in person or electronically via the internet)
- intranet, web or cloud platform – updated documents, policies and procedures
- newsletters – bulletins and updates (hard copy or electronic)
- noticeboards – displaying work plans and information for staff (a free online noticeboard is available at www.padlet.com)
- telephone – a call to check on a team member's progress
- written – letters, memos, reports and minutes (hard copy or electronic).

Example

Imran needs to notify his team quickly of a change to a meeting date. He e-mails and sends a text to all his team members. He also telephones two members who he knows do not have direct access to e-mail or text messages. This way he has used several communication methods to meet the needs of his team.

You will need to develop the skills which enable you to use all the methods of communication which are practical, and to decide which is the most suitable for a particular situation and person. You should consider the advantages and limitations of the different methods and decide if you need any training, for example, using an aspect of new technology.

In some cases, more than one method of communication may be needed. For example, you might have an informal discussion with team members and follow this up with an e-mail to confirm what was decided.

The way you communicate with your team might be influenced by your personality. For example, you might prefer to use e-mails rather than the telephone or text messages. Whichever method you use, you need to make sure that what you convey is understood and acted upon by everyone. You need to be seen as a respected and trusted source of accurate information. You might not be liked by everyone in your team, however you are performing a professional role and you are not there to be everyone's friend. Don't take it personally if you feel someone doesn't like you; it's probably the situation they don't like rather than you as a person.

It's good to be aware of your verbal and non-verbal body language, for example, not folding your arms when speaking as this could look defensive. You also need to take into account the way you speak and act, as your mannerisms might be misinterpreted by others.

Activity

Think back to the last time you communicated with your team members, perhaps during a meeting. How did you act and react to different people and situations? Why was this, and would you do anything differently next time?

Records should always be maintained of all formal communications. This will enable them to be referred to at a later date, for example, if there is any doubt about what was actually said in a meeting, or if actions which should have been completed have not been. Although you might feel it's time consuming, if clear, unbiased, factual and specific records are kept, they should not be misinterpreted.

Informal communications can be more personal and may be quicker if you need to get your team members to react immediately to new information. However, not having a record could prove disadvantageous if a team member has no recollection of you asking them to do something. This would be of particular importance if any issues result in disciplinary action or are relied on during staff appraisals.

Skills of communicating effectively include the way you speak, listen and express yourself, for example, with non-verbal language. Understanding a little about your own personal communication style will help you create a lasting impression upon your team members.

Interpersonal and intrapersonal skills

A way of differentiating between interpersonal and intrapersonal skills is to regard interpersonal skills as *between people* and intrapersonal skills as *being within a person*. Understanding and using these skills will help you develop a range of creative communication techniques appropriate to the activities you require your team members to perform.

Interpersonal skills are about the ability to recognise distinctions between other people, to know their faces and voices, to react appropriately to their needs, to understand their motives, feelings and moods, and to appreciate such perspectives with sensitivity and empathy. Possessing interpersonal skills will help you develop personal and professional relationships.

Ways to improve interpersonal skills include:

- being a mentor to others

- getting organised

- meeting new people at work, social groups, clubs, meetings, etc

- participating in workshops or seminars in interpersonal and communication skills

- spending time each day practising active listening skills with friends, family, colleagues and others

- starting a support/network group in person or online.

Intrapersonal skills are about having the ability to be reflective and access your inner feelings. Having this ability will enable you to recognise and change your own behaviour, build upon your strengths and improve your limitations. This should result in quick improvements, developments and achievements as people have a strong ability to learn from past events and from others.

Ways to improve intrapersonal skills include:

- attending courses, e.g. Neuro Linguistic Programming (NLP), Transactional Analysis (TA) and Emotional Intelligence (EI)

- creating a personal development plan and working towards it

- developing an interest or hobby

- keeping a reflective learning journal

- meditation, or quiet time alone to think and reflect

- observing people who are great leaders, motivators or positive thinkers

- reading self-help books

- setting short- and long-term goals and following these through.

Howard Gardner (1993, p.263) defines intrapersonal intelligence as: *sensitivity to our own feelings, our own wants and fears, our own personal histories, an awareness of our own strengths and weaknesses, plans and goals.* He is best known for his theory of multiple intelligences of which there are eight, interpersonal and intrapersonal being two of them. The other six are:

- linguistic – the ability to use language to codify and remember information; to communicate, explain and convince

- logical – also known as mathematical intelligence; the capacity to perceive sequence, pattern and order, and to use these observations to explain, extrapolate and predict

- musical – the capacity to distinguish the whole realm of sound and, in particular, to discern, appreciate and apply the various aspects of music (pitch, rhythm, timbre and mood), both separately and holistically

- naturalist – the ability to recognise, appreciate and understand the natural world; it involves such capacities as species' discernment and discrimination, the ability to recognise and classify various flora and fauna, and knowledge of and communion with the natural world

- physical – also called kinaesthetic intelligence; the ability to use one's body in highly differentiated and skilled ways, for both goal-oriented and expressive purposes; the capacity to exercise fine and gross motor control

- visual-spatial – the ability to accurately perceive the visual world and to re-create, manipulate and modify aspects of one's perceptions.

According to Gardner, individuals possess all of these intelligences. However, they are not all present in equal proportions (in extreme circumstances it may appear that an individual is severely lacking in one or more). The particular combination of intelligences and their relative strengths can form a profile that is unique to each individual. Some people are more intelligent than their peers; others appear superior at certain tasks, are more capable of manipulating information or more readily see the solutions to problems. Some are more expressive or more capable of learning new tasks quickly.

Gardner's eight intelligences have been debated and amended by many other theorists over time. They can be compared to various learning preference theories, for example, Honey and Mumford's (1986) Activist, Pragmatist, Reflector and Theorist. Being aware of differing intelligences within your team members, and in yourself, will help you consider alternative ways of communicating with your staff.

Extension activity

Consider how you will communicate with your team members. Do you prefer a formal or informal approach and why? Review the theories mentioned so far in this chapter and reflect on how they might influence the way you act and react with your staff.

Updating the work plan

Your original work plan may need updating at some point due to unforeseen circumstances such as a team member leaving or a forthcoming external inspection. All changes will need to be communicated to your team members via the next team meeting or by another quicker appropriate method such as telephone or e-mail.

As a manager you should keep up to date regarding what is happening in your area of responsibility and within your organisation. This also means keeping up to date with changes to the qualification or the programme of learning being assessed and quality assured. If your team is working with accredited or endorsed qualifications, you will need to ensure they all read the latest updates from the awarding organisation. It's useful to sign up for relevant newsletters and check websites regularly. Any developments should be disseminated and discussed at team meetings with minutes documented and distributed.

Examples that might necessitate a change to your work plan could include (in alphabetical order):

- a report requiring immediate action
- an increase or decrease in learners/assessors/quality assurers
- changes to funding
- developments with resources, i.e. new technology
- external inspections
- financial or budget constraints
- organisational developments, i.e. additional locations for assessment
- policy and procedural changes
- qualification or programme changes
- staff turnover
- targets not being met
- updated documentation requiring immediate use.

Activity

What situations might occur within your organisation which would necessitate a change to your work plan? How would you update your work plan, and how would you communicate the changes to your team?

If you are involved in communications from those in higher positions than yourself, you should be kept up to date with what is happening in your organisation. If you don't attend all the meetings or read e-mails and updates you might not know about any important developments and changes. When changes do occur, it is best that you convey them to your team members immediately, rather than have them hear rumours that might not be true, for example, a takeover bid or possible redundancies.

When amending your work plan, if it's a hard copy, try not to use correction fluid, but cross out and rewrite any changes to allow the original information to still be seen. This is useful in case of any queries as to why activities were changed or what the original target dates were. If you are using an electronic work plan you can resave it as a different version with a new date. However, do keep a note of which is the latest version in case you accidentally refer to an outdated one.

Your work plans should be regularly reviewed to make sure that they are still fit for purpose, and updated to meet changes and new circumstances. When amending your plans, you are likely to be more successful if you include your team members in the making of major decisions rather than imposing changes upon them. As their job roles will be affected directly, discussing these and communicating with your team on an ongoing basis could help alleviate future problems.

Example

Celina needed to amend her observation work plan for her team of ten assessors due to her relocation to the organisation's European site for two months. She sent an e-mail to all those whom she was due to observe during that time period, giving them a new date. Almost immediately, she received replies from nearly all her assessors stating the dates were not suitable. On reflection, Celina should have asked her team members first for suitable alternative dates. She would then not have had to deal with several annoyed individuals as well as wasting her time making rearrangements.

Communication is a two-way process and at some point you may have to reallocate certain team members' responsibilities. If you can do this by agreement rather than imposition you should maintain the respect and support of your team. If you do have to impose a change on an individual, try to do so with tact and diplomacy. Tell your team members why you are making the changes and that you need their support. Some people might resist change, as they are comfortable with the way they currently do things. During meetings you could stress that change is inevitable in training and education and that you all need to react to it in a positive way. Treating change as an opportunity for improvement can be enlightening and motivating. Ensuring your team members and yourself remain current with regards to practice should help towards a smoother transition when any changes do take place.

When staff decide to leave your organisation, you will need to plan who will take over their workload and ensure there is a smooth transition so as not to disadvantage anyone. Team members should give a period of notice to allow this to take place. If someone leaves quickly without notice, this will cause a temporary disruption. It must be dealt with in a way that does not increase someone's workload to the extent that they can't cope.

Feedback

You should give informal feedback to your team members as well as formal feedback, on an ongoing basis. Don't think it has to be negative; positive feedback and praise could be given when deserved, for example, if a team member does something exceptionally well. If there is a culture of giving constant feedback, your team members are more likely to be used to it, to listen and to respond to your comments. Feedback should always be constructive and developmental and should be given individually when possible, and to teams when relevant, perhaps during a meeting.

There is a chance that in some companies, feedback is only given in formal situations such as staff appraisals, after an observation or a sampling activity. If this is the case, individuals might come to dread the formal meeting as they may feel they have done something wrong. Giving praise and clarifying situations when you have the opportunity should help motivate your staff and increase morale. Individuals need to know what they are doing right, as well as what they are doing wrong. Creating a culture of giving and receiving regular feedback will help break down barriers, increase motivation and encourage staff to feel valued. This can contribute to positive working relationships and can make more formal feedback sessions, such as staff appraisals, run smoother.

Example

Keiran, an IQA, was walking past a training room and noticed a group of learners who were becoming rather disruptive. He saw Alfons, their assessor, immediately deal with the situation. When Keiran next saw Alfons in the staffroom, he said how pleased he was that the situation had been dealt with in a quick and amicable manner. This left Alfons feeling that what he had done was worthwhile and had been noticed.

According to Kermally (2002) feedback, apart from being ongoing, should also be a development tool. It should allow skills gaps in teams and individuals to be identified, and provide opportunities for the necessary skills to be obtained. Carrying out staff appraisals should be seen as a positive way of discussing a person's job role, any concerns they might have, and identifying any training needs. If action points are agreed, these should always be followed up by both parties. Please see Chapter 2 for more information regarding feedback skills.

Conflict

Conflict can arise when an individual or a group believe that someone has done (or is about to do) something they disagree with. It could be minor, such as an individual sitting in another's usual seat at a meeting. Or it could be more serious, such as a dispute over who carries out certain tasks. There could be challenges and barriers that your staff might face which you weren't aware of, for example, transport issues. This might involve you reallocating staff to make the location more accessible to them. A team member might have a complaint or grievance against another team member and it might be your responsibility to deal with it. You should always base your judgement upon facts after listening to both sides, remain impartial, and be fair and ethical with your decision. Make sure you keep records for future reference in case of any queries.

Mullins (2013) identified a range of potential sources of conflict with individuals, groups and organisations.

Individuals may come into conflict because they have different attitudes, personality styles or particular needs. In some instances, the situation might be aggravated by stress or illness. For example, one team member may perceive quality assurance activities in a different way to another. A reason for different attitudes can be due to the age gaps between different individuals. An older person might feel they have the necessary skills, knowledge and experience, but the power and responsibility might lie with a younger person.

Groups might come into conflict because individuals have different skills, attitudes and ways of working. Team members might interpret tasks in different ways, new staff might feel excluded or a key person might be absent. Some characteristics can create conflict, for example, the hierarchy structure, management or leadership styles. This could result in differences and disagreements between departments or teams.

Sometimes, simple situations can easily lead to conflict if they are not dealt with immediately or are misinterpreted. Ongoing communication and feedback should help confirm your expectations of your team members, and their expectations of you. Always establish the root cause of any situation that leads to conflict, to work out a strategy to resolve it. Opportunities should be taken to clarify any misunderstandings and the whole team should be informed to ensure everyone is working to the same ethos. Allowing conflict to worsen over time can lead to a situation becoming much more difficult to resolve. Conflict can also affect others who are not directly involved in the original situation. For example, if you don't take any action, then your own manager might perceive you as being ineffective. This could affect the confidence they have in you as a manager, and it could also affect the respect you have from your team members.

Extension activity

Refer to your original work plans created earlier on in this chapter. What changes would you now make to them and why? What developments might occur in the future which would require you to amend them? Will you act differently with your team members now that you have gained further knowledge of communication techniques?

Summary

Knowing how to plan, allocate and monitor the work of others should help you become more productive in your role. Understanding various management theories should help you communicate effectively with your team.

You might like to carry out further research by accessing the books and websites listed at the end of this chapter.

This chapter has covered the following topics:

- Producing and using a work plan

- Identifying and allocating responsibilities to team members

- Monitoring the progress of others and the quality of their work

- Communication skills

- Updating the work plan

References and further information

Adair, J. (2002) *John Adair's 100 Greatest Ideas for Effective Leadership and Management*. Mankato: Capstone.

Armstrong, M. (2003) *A Handbook of Human Resource Management*. London: Kogan Page.

Armstrong, M. (2008) *How to Be an Even Better Manager* (7th edition). London: Kogan Page.

Bacal, R. (2004) *How to Manage Performance: 24 Lessons for Improving Performance*. New York: McGraw-Hill.

Belbin, M. (2010) *Team Roles at Work* (2nd edition). Oxford: Butterworth-Heinemann.

Berne, E. (1973 and 2010 editions) *Games People Play: The Psychology of Human Relationships*. London: Penguin Books.

Coverdale, R. (1977) *Risk Thinking*. Bradford: The Coverdale Organisation.

Friedman, M. and Rosenman, R. (1974) *Type A Behaviour and Your Heart*. New York: Random House.

Gardner, H. (1993 and 2011 editions) *Frames of Mind: Theory of Multiple Intelligences*. New York: Basic Books.

Handy, C. and Constable, J. (1988) *The Making of Managers*. London: Longman.

Honey, P. and Mumford, A. (1986 and 1992 editions) *Manual of Learning Styles*. Coventry: Peter Honey Publications.

Kennedy, C. (2007) *Guide to the Management Gurus* (5th edition). London: Random House.

Kermally, S. (2002) Appraising Employee Performance. *Professional Manager*, 11(4): 30–1.

Mullins, L .J. (2013) *Management and Organisational Behaviour* (7th edition). London: Prentice Hall.

Rosenman, R. H., Brand, R. J., Sholtz, R. I. and Friedman, M. (1976) Multivariate prediction of coronary heart disease during 8.5 year follow-up in the Western Collaborative Group Study. *The American Journal of Cardiology, 37*(6): 903–10.

Wallace, S. and Gravells, J. (2007) *Leadership and Leading Teams*. Exeter: Learning Matters.

Wallace, S. and Gravells, J. (2007) *Mentoring*. Exeter: Learning Matters.

Websites

360 degree feedback – www.cipd.co.uk/hr-resources/factsheets/360-degree-feedback.aspx

Carl Jung – www.simplypsychology.org/carl-jung.html

Friedman and Rosenman Type A and B personalities – www.simplypsychology.org/personality-a.html

Gantt charts – www.mindtools.com/pages/article/newPPM_03.htm

Institute of Leadership and Management – www.i-l-m.com

Kiersey temperament theory – www.kiersey.com

Myers Briggs type indicator – www.myersbriggs.org

Online noticeboard – www.padlet.com

SWOT analysis – www.businessballs.com/swotanalysisfreetemplate.htm

Tuckman – www.infed.org/thinkers/tuckman.htm

Work plans – http://cec.vcn.bc.ca/cmp/modules/pm-pln.htm

CQFW	Credit and Qualification Framework for Wales
CRB	Criminal Records Bureau (now part of DBS)
DBS	Disclosure and Barring Service
DCELLS	Department for Children, Education, Lifelong Learning and Skills (Wales)
DfE	Department for Education
DSO	Designated Safeguarding Officer
E&D	Equality and Diversity
EBD	Emotional and Behavioural Difficulties
ECDL	European Computer Driving Licence
EDAR	Experience, Describe, Analyse, Revise
EDIP	Explain, Demonstrate, Imitate, Practise
EHRC	Equality and Human Rights Commission
EI	Emotional Intelligence
EQA	External Quality Assurance/Assurer
ESOL	English for Speakers of Other Languages
ETF	Education and Training Foundation
EV	External Verifier
FAQ	Frequently Asked Questions
FE	Further Education
FELTAG	Further Education Learning Technology Action Group
FHE	Further and Higher Education
GCSE	General Certificate of Secondary Education
GLH	Guided Learning Hours
H&S	Health and Safety
HEA	Higher Education Academy
HEI	Higher Education Institution
IAG	Information, Advice and Guidance
IAP	Individual Action Plan
ICT	Information and Communication Technology
IfA	Institute for Apprenticeships
IfL	Institute for Learning (no longer operational)
IIP	Investors In People
ILA	Individual Learning Account

ILM	Institute for Leadership and Management
ILP	Individual Learning Plan
ILT	Information and Learning Technology
IM	Internal Moderator
IQ	Intelligence Quotient
IQA	Internal Quality Assurance/Assurer
ISA	Independent Safeguarding Authority (now part of DBS)
IT	Information Technology
ITE	Initial Teacher Education
ITOL	Institute of Training and Occupational Learning
ITP	Independent Training Provider
ITT	Initial Teacher/Trainer Training
IV	Internal Verifier
IWB	Interactive Whiteboard
LA	Local Authority
LAR	Learner Achievement Record
LDD	Learning Difficulties and/or Disabilities
LLUK	Lifelong Learning UK (no longer operational)
LRC	Learning Resource Centre
LSA	Learner (or Learning) Support Assistant
LSCB	Local Safeguarding Children Board
LSIS	Learning and Skills Improvement Service (no longer operational)
LWI	Learning and Work Institute
MLD	Moderate Learning Difficulties
MOOCs	Massive Open Online Courses
NEET	Not in Education, Employment or Training
NIACE	National Institute of Adult Continuing Education (now the LWI)
NLH	Notional Learning Hours
NLP	Neuro Linguistic Programming
NOS	National Occupational Standards
NQT	Newly Qualified Teacher
NRDC	National Research and Development Centre for adult literacy and numeracy
NTA	Non-Teaching Assistant

NVQ	National Vocational Qualification
OER	Open Education Resources
Ofqual	Office of Qualifications and Examinations Regulation
Ofsted	Office for Standards in Education, Children's Services and Skills
OU	Open University
PAT	Portable Appliance Testing
PCET	Post Compulsory Education and Training
PDBW	Personal Development, Behaviour and Welfare
PGCE	Post Graduate Certificate in Education
PLTS	Personal Learning and Thinking Skills
POCA	Protection of Children Act 1999
PPE	Personal Protective Equipment
PPP	Pose, Pause, Pick
PSHE	Personal, Social and Health Education
QCF	Qualifications and Credit Framework (replaced with RQF)
QSR	Qualification Success Rates
QTLS	Qualified Teacher Learning and Skills (further education and skills)
QTS	Qualified Teacher Status (schools)
RARPA	Recognising and Recording Progress and Achievement in non-accredited learning
RIDDOR	Reporting of Injuries, Diseases and Dangerous Occurrences Regulations
RLJ	Reflective Learning Journal
RPL	Recognition of Prior Learning
RQF	Regulated Qualifications Framework (replacement to QCF)
RWE	Realistic Working Environment
SAR	Self-Assessment Report
SCN	Scottish Candidate Number
SCQF	Scottish Credit and Qualifications Framework
SET	Society for Education and Training
SfA	Skills Funding Agency
SL	Student Loan
SLC	Subject Learning Coach
SMART	Specific, Measurable, Achievable, Relevant and Timebound

SoW	Scheme of Work
SP	Session Plan
SQA	Scottish Qualifications Authority
SSB	Standard Setting Body
SSC	Sector Skills Council
SWOT	Strengths, Weaknesses, Opportunities and Threats
T&L	Teaching and Learning
TAQA	Training, Assessment and Quality Assurance
TNA	Training Needs Analysis
UCU	University and College Union
ULN	Unique Learner Number
VACSR	Valid, Authentic, Current, Sufficient, Reliable
VARK	Visual, Aural, Read/write, Kinaesthetic
VB	Vetting and Barring
VET	Vocational Education and Training
VLE	Virtual Learning Environment
WBL	Work Based Learning
WEA	Workers' Educational Association
WWWWWH	Who, What, When, Where, Why and How

INDEX